Eight Hundred Years of Physics Teaching

George Bishop

Dawn Bishop.

Fisher Miller
PUBLISHING

First published 1994

Published by Fisher Miller Publishing
Wits End, 11 Ramsholt Close
North Waltham, Basingstoke
Hampshire
RG25 2DG
United Kingdom.

Printed by Turnergraphic Ltd, Winchester Road, Basingstoke

ISBN 1-899077-00 6

A catalogue record for this book is available from the British Library

Acknowledgements
The cover photograph appears by kind permission of the Royal Institution, 21 Albermarle Street, London.

Table of Contents

Appendices

Introduction

Eight hundred years of physics teaching

Many readers might express surprise at such a title: eight hundred years? Surely physics teaching only began one or two centuries ago with Newton and Faraday and the like?

That was my conception too, until I began probing back; but the more I probed the further back in time I went. I have taken the start of physics teaching 'proper' in Europe to be the renaissance of learning in the twelfth century. It was then that the scientific works of the Greeks, the Latins and the Arabs were received into Europe; it was then that the first 'studia generalia' or universities were founded; it was then that the Franciscans and Dominicans and Benedictines established their houses of learning.

One might be tempted to discount the early beginnings of physics and of its teaching as being too elementary and insignificant. But it was on these humble foundations that the giant edifice of science was built. The nascent science of each age was pursued with a keen curiosity, with apparatus as efficient as was then procurable. It was no less a genius than Sir Isaac Newton who said, 'if I have seen further it is by standing on the shoulders of giants'. T. Martin, in his history of the Royal Institution, wrote:

> History is, and must always largely be, the story of the great ones among us, those who have guided and influenced the activities of their fellow men.[1]

Similarly, Emerson claimed: 'There is properly no history; only biography'.

As the chapter headings show, this history of physics teaching is structured on a biographical pattern. Whenever possible the story of an influential person or institution is used to tell the story of the period. The book begins with the contributions of the Greeks, the Romans and the Arabs to physics and its teaching. It was the introduction of their works into northern Europe which brought about that great upsurge of learning which, in turn, prepared the soil and sowed the seeds in the universities of Oxford and Cambridge and Paris for that flowering of 'modern' physics exemplified in the works of Sir Isaac Newton and others.

I often wondered how and where such great geniuses as Isaac Newton, Faraday and others were taught their physics, and by whom, in those far-off days 300 years ago. Chapter 2 throws light on Newton's education in physics. Not only did Newton learn his physics at the University of Cambridge: his own discoveries were to contribute much to the teaching of the subject, not just at Cambridge, but also at Oxford and the Scottish and other European universities. With the onset of the Industrial Revolution the universities no longer remained the sole preserves of science and of its teaching. An appreciation among the rising new industrial and middle classes of the need for more enlightened curricula and pedagogy (embracing, for example, more experimental methods in the teaching of the sciences) became widespread. It was in response to these demands that numerous academies and other schools and educational establishments that specialized in mathematics and physics teaching arose during the latter half of the 17th and during the 18th centuries.

It was in private academies and schools such as these that many great men of science – such as Joseph Priestley, for example – received their education (Chapter 3). These dissenting (non-conformist) and similar academies catered largely for the middle classes. It was to bring science within the reach of the working classes that Mechanics' Institutions were established throughout the country. Chapter 4 considers the work of men such as George Birkbeck and others who were instrumental in taking physics education to the masses. The College named after Birkbeck, which today is a constituent part of the University of London, stands as a memorial to the efforts made on behalf of the education of the less affluent members of society.

Chapter 5 deals with the Royal Institution and Michael Faraday. The two are inseparable. The Royal Institution was one of many scientific societies and institutions founded in the 18th and 19th centuries to further the cause of science and its teaching. The Institution has been served by a long line of scientists who were not only brilliant in their own fields but were also gifted lecturers. We may not be able to reach the dizzy heights of scientific discovery reached by Faraday, but we can all learn from the example he set us. He made a special study of the art of lecturing, an example present-day lecturers could follow with profit, not only to themselves but also to their audiences. The chapter considers not only the work of Faraday but also that of other eminent Professors of the Institution, such as Davy, his predecessor there, Thomas ('Phenomenon') Young, John Tyndall, James Dewar, Lord Rayleigh, J.J. Thomson, Ernest Rutherford and the father and son William and Lawrence Bragg.

Chapter 6 deals with the beginnings of active involvement by the state in elementary education in the country. It was the church and philanthropic institutions that were the first into this field. The simple science taught in their schools consisted mainly of oral, conversational lessons or object lessons. With the large numbers in the classes of the monitorial schools any other method of teaching would have been difficult. But there were other schools where more enlightened methods prevailed. One such was the school set up by the clergyman Richard Dawes at King's Somborne in Hampshire, where children were taught science practically, using simple apparatus, and concentrating on matters that were familiar and interesting and of future use to the pupils. There *can* be no effective science teaching without knowledgeable and enlightened teachers. To remedy this deficiency Dawes wrote excellent manuals of science teaching, using a wealth of practical examples. Associated with Dawes was another clergyman, Henry Moseley. Moseley was the first school inspector ever to have an academic background in science. He drew up lists for the Government (the Science and Art Department) of apparatus suitable for schools and for teacher training colleges.

Chapter 7 looks at secondary education in the first half of the 19th century. Most of the public and the grammar schools remained hidebound by the tradition of the classics. But many private schools and academies sprang up to provide an education more in tune with the demands of the new commercial and industrial age. Many of these were well equipped with apparatus which was put to effective use. Among these schools were the two attached to University College and King's College, two recently founded constituent colleges of the University of London. Indeed, it was a Professor of Greek at University College – Henry Malden – who, in a lecture in October 1838, gave as excellent a description of how science should be taught as any that could be given by pedagogues of today. The Matriculation Examination of the University of London, for which physics was a compulsory subject of study, did much to stimulate study of the subject throughout the country. The Department of Science and Art, established in 1853, provided a great stimulus to the teaching of the elements of science in the second half of the 19th century.

Chapter 8 considers its many schemes – such as 'payment by results' – that were to bring this about. There was an urgent need for an increased knowledge of science in the population. The Department decided it would pay those teachers and institutions who would bring this about. Incentives were needed if science was to get into the schools. But such incentives received a setback, when, in the interests of economy, the Revised Code of 1861 paid grants only for the three Rs and religious knowledge. Very active in getting science into the schools was the towering person of T.H. Huxley. He was an extremely influential figure on the London School Board, which was to set an example for science teaching to the many school boards that were set up in towns and cities throughout the land. The London Board provided peripatetic demonstrators who visited local schools, carrying their own apparatus with them. In order to receive the grants of the Science and Art Department many elementary and private and grammar schools accepted its condition, namely, to provide regular instruction in science over three years. So came into being the 'organized science schools'.

Meanwhile the great public schools of Britain remained bogged down in tradition and the classics (Chapter 9). Between 1861 and 1864 the Clarendon Commission carried out an inspection of all public schools. The resulting Public Schools Act of 1868 freed the public schools from their own shackles, imposed by tradition and long out-of-date statutes, and science came to be introduced into Rugby and Eton.

Long before this, though, the 'lesser' public schools, such as Stonyhurst College in Lancashire and University College School, London, were teaching physics, making use of laboratory apparatus and equipment (Chapter 10). When the Schools Inquiry Commission (1864–8) looked into secondary education (excluding the public schools) it found that in very few cases was effective instruction given in science.

Some private and proprietary schools which did give such instruction, for example, Queenswood College in Hampshire, were the rare exceptions (Chapter 11). The Royal Commission on Scientific Instruction and Advancement of Science confirmed (1875) that little or no science was taught in the grammar schools. The Commission, after disposing of the usual arguments brought against introducing science into the schools – lack of funds for special teachers and apparatus, overloading of the curriculum and so on – recommended that not less than six hours a week should be devoted to science in all schools. It gave a further fillip to physics teaching in the schools by making suggestions as to what physics should be taught and how. In 1895 the Bryce Commission reported that there was too much science in several grammar schools, to the detriment of literature and the arts.

Chapter 12 considers the role and influence of external examinations. Examinations always have exerted (and still do exert) great influence on what goes on in the schools. As stated earlier, the University of London Matriculation Examination, in which physics was a compulsory subject, played a major role in ensuring that physics secured a prominent place in the curriculum of schools. The Oxford and Cambridge Local Examinations also became important levers, not only in introducing science into the schools, but in influencing women's education throughout the country. For decades all teaching was based on the lecture/didactic method. In due course there was a reaction against this 'all teacher, no pupil' approach. Out of this reaction the heuristic (discovery) method of science teaching was born, the essence of which is 'all pupil, no teacher'.

Chapter 13 examines H.E. Armstrong's heuristic method, a forerunner of the later Nuffield approach, and its influence on science teaching in schools.

Chapter 14 looks at the state of physics teaching in the universities during the 19th century. At both Oxford and Cambridge there were demands for more and better physics teaching at these universities. The University Reform Acts of 1864 (Oxford) and 1866 (Cambridge) instituted reforms which were to transform the two universities into the world's leading centres of learning. The establishment of the Clarendon Laboratory at Oxford and of the Cavendish at Cambridge were significant events in raising the status and popularity of physics. At the time, a knowledge of physics was demanded of all candidates for a general degree – a far more enlightened approach than applies today. The colleges each had their own tutors, many of whom wrote textbooks in response to demand for them. The high standard of physics teaching in the Scottish universities was not only maintained but raised in the 19th century in consequence of a long line of brilliant men who held the Chairs of Natural Philosophy: men such as William Thomson (Lord Kelvin) at Glasgow and James Clerk Maxwell at Aberdeen. Maxwell was also Professor of Physics at King's College, London. The chapter considers developments in physics teaching not only at London University, but also in the nascent provincial universities. It has been rightly said that at the Cavendish Laboratory in Cambridge more physics was done per square centimetre than in any other laboratory in the world: understandably it was the Mecca of all physicists.

Chapter 15 looks at the Cavendish under the stewardship of such famous physicists as James Clerk Maxwell, Lord Rayleigh, J.J. Thomson (later Lord Kelvin), William Bragg and Ernest Rutherford. It was at the Cavendish that virtually all the major discoveries and new developments in physics were made during the 19th and early 20th centuries. During the early part of the 20th century the system of grants from the Government ensured that science figured in the school curriculum. In the elementary and lower secondary school this science consisted of object lessons, with practical work consisting of an endless round of measuring. In the Higher Grade and Central schools the practical work was a bit more interesting – making electric bells, for example. Here Armstrong's heuristic method prevailed (Chapter 16).

There were several attempts to make science a compulsory subject for all pupils; it was usually an alternative to mathematics. The most important change in science teaching in schools in the first half of the century was the development of the General Science movement. There was a widespread demand for a broader science education to form part of a general education of every boy and girl. By the end of the

1920s heurism had run out of steam. However, it had paved the way for the next big innovation in science teaching – the Nuffield approach, with its emphasis on enquiry. The Nuffield approach to science teaching profoundly changed the whole nature of physics education in the schools. 'Science for the enquiring mind' will no doubt continue through the 1990s and beyond, but supplemented with consideration of the interrelationships between science, society and the environment.

Chapter One

The First Beginnings of Physics Teaching

1. The Greeks

The development and study of physics as a science began with the Greeks. In the fields of statics and mechanics the Greeks were acquainted with the principles of the lever, the rudder, the wheel and axle, the balance, the beam, the screw, and the wedge.

Aristotle and Archimedes

The influence of Aristotle (384–322BC) who had grasped the notion of the parallelogram of forces, on science and science teaching was incalculable almost to the end of the 17th century. For almost 2000 years his works remained one of the few standard authorities on science. Unfortunately the physical treatises of Aristotle were not as promising as those of his other scientific works. He believed a vacuum to be impossible. He asserted that motion could only be sustained by continuous application of force, and that bodies fall quicker in exact proportion to their weight.

Archimedes (c.287–212BC) discovered the principle of the compound pulley and of the endless screw. Apart from his work on mechanics and statics, Archimedes greatest contributions were made in the science of hydrostatics. His 'Principle' is familar to the modern schoolchild. The Greeks were acquainted with the principles of the siphon and of the pump. They showed great adeptness in the construction of 'automatic' hydraulic contrivances. Besides Archimedes, the science of hydrostatics was furthered by the writings of Hero, the inventor of 'Hero's engine', an aeolopile or simple reaction steam turbine. Hero wrote a number of semi-popular books on mathematics, mechanics and pneumatics. In optics the Greeks were acquainted with the rectilinear propagation of light, the laws of reflection, and the phenomena of refraction. They attempted an explanation of the rainbow.

Euclid, Ptolemy and Pythagoras

The fundamental principles of geometrical optics were laid down by Euclid (fl.323–285BC), as were the properties of convex and concave mirrors. Ptolemy (d.AD127–157) also did good work in the study of light. The *Optics*, ascribed to him, contains an experimental study of refraction which is one of the most remarkable experimental investigations of antiquity. He measured the angles of incidence and of refraction and arranged them in tables. In the science of magnetism and electricity the Greeks were aware that lodestone attracted iron and that amber, when rubbed, would attract light bodies. Pythagoras (d.497/6BC) laid the foundation of the mathematical study of sound. The Greeks were aware that sound was transmitted by vibrations in the air. But all this science was a closed book to the rest of Europe.

2. The Dark and Early Middle Ages

During this period the development of physics and of physics education was extremely slow. After the arrival in Britain of St Augustine in AD596 schools were established by the great monastic churches, for example at Canterbury, York and Westminster. These schools provided instruction in Latin language, in the Scriptures, and in Church liturgy and music. The early Christian fathers were not interested in science. Their interest lay in religion, in things of heaven rather than of earth. The chief civilizing influences in Europe between the sixth and ninth centuries were the Benedictine monasteries and schools. Attention was possibly given to some topics of elementary physics in the courses comprising the seven liberal arts. The earliest physics teaching was probably given by men such as Venerable Bede (673–735) and his monks. In his *De Natura Rerum* Bede continually refers the phenomena of nature to natural laws and causes. He taught that lightning and thunder were due to collision of clouds. He was acquainted with the properties of lodestone. However, what little physics teaching there was, was given in a manner quite different from that today. The interest of the monks in physics was philosophical, not practical; physics constituted a branch of philosophy. In an agricultural, feudal society, in an age of little machinery, there was no need for the 'practice' of physics. During the 11th century a more active desire for a more secular knowledge appeared, arising from increasing commerce, urbanization and contact with new peoples and new ideas.

3. Arabic Science

By the ninth century AD the excellence of Muslim science was well established, and Arabic became the international vehicle of scientific progress till the 12th century.

The best known of the Muslim physicists were Al-Kindi (d. c.873), Alhazen (c. 965–1039), Al-Biruni and Avicenna (980–?). Al-Kindi wrote a treatise on geometrical and physiological optics based largely on Euclid, Hero and Ptolemy.

Ibn al-Haitham, or Alhazen, of Cairo, was probably the greatest Arabic physicist of medieval times. The Latin translations of his works such as the *Kitab al Manazir* remained the standard textbooks on optics in Western Europe until the 17th century. In the 1490 Statutes of Oxford University the *Optics* of Alhazen was one of the recommended books. By his research in catoptrics he worked out the theory of spherical and parabolic mirrors, and spherical aberration. By his work on dioptrics he discovered that the ratio between the angles of incidence and of refraction does not remain constant. He did much work on the magnifying power of a lens. He determined the focus of reflected rays when the position of the eye and of the object were known. He gave the correct explanation of the apparent increase in the size of the moon and of the sun when near the horizon. He gave a good anatomical description of the eye. He taught that the rays of light irradiate in the object seen and not in the eye. The earliest use of the camera obscura is found in his work. Alhazen also dealt with other topics of optics, such as shadows and colours, optical illusions, mirages, haloes, comets, the Milky Way, and the rainbow.

Al-Biruni, in addition to his fine work on specific gravity, observed that the speed of light was incomparably greater than that of sound. He explained the working of natural springs and artesian wells by the hydrostatic principle of communicating vessels. Ibn-Sina (Avicenna) investigated all the fundamental questions of physics of his time, e.g. motion, contact, force, the vacuum, light, heat, specific gravity, music. His writings played an important role in medieval learning.

In the 12th century the Muslim world gave to medieval Europe one of the most remarkable treatises on science. This was the *Book of the Balance of Wisdom* (1121) by Al-Khazini, which dealt at length with several problems of mechanics, such as the application of the balance to levelling and to the measuring of time, the principle of the lever, the theory of gravity – a universal force directed to the centre of the earth, the weight of the air, and so on. Concerning hydrostatics, he gave the specific gravities of many liquids and solids, and he discussed the use of an aerometer to measure the densities and temperatures of liquids. He also made careful observations on capillarity which show that he had some empirical knowledge of surface tension.

4. The 12th-century Renaissance

During the 12th century there occurred a brilliant renaissance, not only of learning in general, but also of science. This resulted from the reception into Europe of almost the entire works on science of Aristotle and of other Greek scientists, along with the treasures of Arabic science. The great increase in knowledge that resulted from the translation into Latin of such works on physics as those of Ptolemy, Euclid, Hero, Archimedes and Aristotle, along with the scientific contributions of the world of Islam, such as the treatises on optics of Al-Kindi, Alhazen, Avicenna, etc., brought back to Europe by the Crusaders and by wandering scholars, resulted in considerable expansion of the subject matter of the traditional trivium and quadrivium. Evidence of this expansion is provided by the cathedral school at Hereford where the studies included not only the seven liberal arts and theology but also the sciences of astronomy and of physics.

Peter of Blois, who was on intimate terms with both King Henry II of England and William II of Sicily (which was then a flourishing centre of Moorish science), in a letter written in about 1160, not only provides evidence that physics was being taught but criticizes this teaching on the grounds that it is given too soon to boys too immature to appreciate the technicalities. He complains:

> Some before they are imbued with elementary disciplines are taught to inquire about ... matter and motion, the principles of bodies ... what time is, what empty space ... of the divisible and indivisible ... of the causes of things, of the tides of ocean ... of various secrets of latent Nature ... and many other matters which require a foundation of fuller science and more eminent intellects.[1]

Contemporary with this reception of Greek, Latin and Arabic science was the rise of the *studia generalia* or universities of Europe. It was in these that most of the scientific activity of the time took place and the first beginnings of systematic instruction in physics were made. At first they were collections of teachers and pupils rather than organized institutions. Legal privileges were conferred on Cambridge in 1231 and on Oxford in 1241. The 'universities' owned no property: lectures were given in houses, and disputations held in churches. It was not until the 14th century that universities began to erect buildings for their own use, and 'colleges' were established. By 1300 Oxford and Cambridge had between them some 2000 students.

Grosseteste and Bacon

The chief educators at this time were the Franciscans and the Dominicans. During this period the Franciscan friar Roger Bacon wrote his eulogy on experimental science. Another to insist on the importance of mathematical studies and on the value of experimentation was Robert Grosseteste, later Bishop of Lincoln. The numerous allusions in the various university statutes to the study of natural philosophy (i.e. physics) and the recommendations to students to study authors who had written standard textbooks of physics indicate that the study of physics is older than generally believed. Besides recommending the treatises on physics of Aristotle, the references to Euclid, Ptolemy, to John Peckham's *Perspectiva Communis* and to the optics of Alhazen and of Witelo, give some idea of the extent of this physics teaching.

The Franciscan Robert Grosseteste (c.1168–1253), was educated at Oxford and Paris universities. Shortly after the resumption of lectures in Oxford he was elected 'Rector Scholarum' and later Chancellor of the University. He introduced Latin translations of the Greek and Arabic scientific writings into the university. He consistently searched for a rational scheme of things, both natural and divine. He was deeply interested in scientific method which he described as both inductive and deductive. He taught that by observation of individual events in nature man advances to a general law, called a 'universall experimental principle' which accounts for these events. Experimentation either verifies or falsifies a theory by testing its consequences. As Crombie says, 'by relating these logical methods to scientific practice Grosseteste made the first moves towards the creation of modern experimental science.'[2] For him the study of nature was impossible without mathematics. He took a special interest in the science of optics and understood perfectly the principles of both the microscope and the telescope.[3] In his study of the rainbow and of comets he used both observation and mathematics.

Grosseteste gave many instructions for the course of study at Oxford, and since his interests lay in the direction of mathematical and physical science the study of physics – at least certain branches of it – was encouraged. It was due primarily to him that lectures on optics, for instance, were given at Oxford.

Crombie writes of Grosseteste:

> For a long period he was in a position to exert a decisive influence over the intellectual interests of the young University of Oxford and he became the founder of a distinguished school of scientific thought. After him there developed at Oxford a characteristic interest both in the methods and implications of

experimental investigation and scientific explanation in general and in the special and experimental and mathematical sciences, a tradition that may be traced in through the works of Roger Bacon, John Peckham, Duns Scotus, William of Ockham, Thomas Bradwardine, John of Dumbleton, and others ... With Grosseteste, Oxford became the first centre of the methodological revolution with which modern science began.[4]

His insistence on the use of experiment or experience and of mathematics ... and the particular significance he gave to the study of optics, remained characteristic of the Oxford school during the 13th and 14th centuries.[5]

One of Grosseteste's pupils was Roger Bacon (1214–92). For Bacon, too, mathematical and experimental methods were the key to science. Man acquires knowledge through reasoning and experiment, but without the latter can have no certainty. Bacon had the added advantage of being able to make use of new sources of scientific knowledge, as, for instance, the works of Al-Kindi and Alhazen.

Bacon made many optical experiments and it seems he was acquainted with the principles of the telescope and of the compound telescope.[6] His treatises on optics and mathematics were probably the most popular science textbooks in the Middle Ages, and he himself gave advanced instruction in these subjects at Oxford.

Bacon in his turn exerted influence on John Peckham (c.1230–1292), whose *Perspectiva Communis* was to remain a classic textbook of optics for nearly 300 years. The references to it in the disputations at Oxford show that the work was well studied. Several manuscripts of the *Perspectiva*, some of which are 'school copies', still exist. The textbook deals with the anatomy of the eye, the physiology of vision, the theory of concave mirrors, the theory of convex and concave lenses, and the rainbow.

Another to be influenced alike by Grosseteste and by Bacon was the Silesian writer and experimenter on optics, Witelo (born about 1230). Around 1270 Witelo prepared a most comprehensive and systematic work on optics, based on Alhazen, but which also contained his own experiments on refraction. In his work Witelo speaks often of experimenting with instruments and repeatedly emphasizes the experimental nature of his subject. He encouraged his readers to experiment for themselves: 'Experience more than books will teach the varied possibilities of images' from mirrors. Witelo's book remained the best and most popular of the textbooks on optics right up till the 17th century, even Kepler paying much attention to it. The 1490 statutes of Oxford University lists Witelo's *Optics* as one of the recommended books for students preparing for the licence to teach, that is for the MA.

Apart from optics, another branch of physics that was taught in the medieval universities was magnetism. In 1269 the Picard Crusader Petrus Peregrinus produced his *Epistola de Magnete,* in which he constantly uses and encourages the experimental method. In this, the first scientific treatise on the subject, Peregrinus describes experiments dealing with the laws of polarity, the magnetization of iron by stroking with a magnet, demagnetization by 'violence', the exertion of magnetic force through water, glass, etc. He describes the construction of a magnetic terella and of an improved floating compass. He also describes a perpetual-motion machine based on magnetic attraction. Roger Bacon, who was probably taught by Petrus, speaks of him in his *Opus Majus* as the greatest experimental scientist of his time.

From the number of extant copies of Petrus's work on magnetism in Oxford, the subject was evidently familiar to Oxford students of the time. The *De Magnete* was not surpassed until the publication of Gilbert's work on the magnet in the 16th century.

In addition to the advances made in optics and magnetism, the 13th century saw the laying of those foundations in statics and dynamics that came to fruition later in the discourses of Galileo and Newton. The founder of the new school of mechanics was Jordanus Nemorarius of Saxony (fl.1222), the second Master-General of the Order of Preachers. Nemorarius visited Oxford during 1229–30. In his treatise *On Weights* is contained the fundamental notion of statical moment and its application to the study of the angular lever and the inclined plane.

Practical studies

In spite of the efforts of Grosseteste and Bacon and others to introduce more experimental physics into the university curriculum, science continued to be taught on Aristotelian lines. The subject, which consisted mainly of Aristotle's physics, was taught dialectically rather than experimentally, by means of lectures and dictated notes, out of libraries, not laboratories.

Numerous writers produced commentaries on the physics of Aristotle. Among these were Simon of Faversham (Simon Anglicus, d.1306) who was Chancellor of Oxford in 1304; Thomas Jorz (Thomas Anglicus d.1310), a Dominican who taught in Paris, Oxford and London; John the Canon, a Franciscan who taught in Oxford around 1329; Walter Burly (1275–c.1335) of Merton College and later tutor to the Black Prince; and William of Collingham. For the benefit of students some writers simplified and condensed the writings of Aristotle and other natural philosophers into 'cram' books, 'so that those matters which were previously drawn out may be readily comprehended under the form of a brief conpendium'.

However, in view of the experimental nature of the work done by men such as Grosseteste, Roger Bacon, and Witelo in optics, by Petrus Peregrinus in magnetism, by Nemorarius and others in mechanics, it is hard to believe that they would not have used some sort of practical demonstration to illustrate their points to students. The only apparatus needed would be such simple items as levers, wedges, pulleys, pumps, balances, mirrors, lenses, magnets, etc. Some, at least, of their students may have been given the opportunity of using such simple apparatus.

5. The 14th Century

It was in the 14th century that the early foundations in statics and mechanics were consolidated to culminate later in the great advances of Galileo and Newton. The chief centres where this mechanics was studied were Merton College, Oxford and Paris University. Scholars who were teaching and pursuing research into mechanics and statics at this time included Bradwardine, Swineshead, John of Dumbleton, William Heytesbury and Jean Buridain.

Thomas Bradwardine

Chiefly responsible for the great activity in mechanics at Merton College was Thomas Bradwardine (c.1290–1349). Bradwardine, who was himself educated at

Merton, lectured there during 1325–35. In 1349 he was elected Archbishop of Canterbury. In 1328 Bradwardine wrote *Tractatus Proportiorum*, a summary of the dynamics and kinematics of the time, which remained a standard textbook till the end of the Renaissance. In another work, *Tractatus de Continuo*, Bradwardine helped lay the foundations of the calculus and of the use of mathematical functions in physics.

Richard Swineshead

Another celebrated textbook on mechanics was the *Calculationes* written by the Cistercian Richard Swineshead (or Suiseth, fl.1350), who had also been educated at Merton. In this he covered the topics of proportions of velocities, acceleration, reaction, maxima and minima, refraction and density, and the intensity of luminous bodies. In his attempts to express in words what was later put more succinctly in symbols and equations, he helped lay the foundations for the later development of the mathematical treatment of physics. The famous 16th-century mathematician, Cardan, included Swineshead among the 12 greatest thinkers of all time.

John of Dumbleton

John of Dumbleton (fl.1331–49) was a Fellow both of Merton and of Queen's colleges. He showed that the space traversed in a given time by a body moving with uniform acceleration was equal to that traversed in the same time by a body moving uniformly with the velocity reached at the midpoint of the time. In the 17th century this rule was to become the basis of the kinematic law of freely falling bodies. The use of graphs to express functional relationships in mechanics and in optics by Dumbleton and by Nicole Oresme (c.1328–82) in dynamics prepared the way for the creation of the algebra and geometry of change that were to transform physics in the 17th century.

William Heytesbury

Heytesbury (fl.1330–71) was a Fellow of Merton College in 1330 and Chancellor of Oxford University in 1371. In his treatises on physics he adumbrated the principles of conservation of mass and of energy. He observed that the path of a falling body is three times longer during the second second of fall than during the first.

Jean Buridain

Buridain (d.1358), who lectured at Paris University, where many English students studied, emphasized that the impetus of a body was directly proportional to its speed, volume and density and described it much as Newton later described quantity of motion (momentum) as mass multiplied by volume. He applied the impetus theory to explain the acceleration of falling bodies.

A 1366 Statute of Paris University for the licence to teach (BA) stipulates under natural philosophy the works of Aristotle, e.g. the *Physica, De Coelo* etc. But an extension of the course in science is evidenced by the requirement that the candidate was in addition 'to have heard other mathematical works' such as John Peckham's textbook on optics.[7] Even earlier, in 1303, Cambridge introduced a new statute in which it was said to be 'not improper to give support to mathematical sciences, including optics, astronomy, now somewhat in danger'.[8]

Marsilius van Inghen, whose ideas were similar to the impetus theory propounded by Jean Buridain, was proctor of the English 'nation' at Paris University from 1373 to 1375. It is said that special arrangements had to be taken to accommodate the students who flocked to hear him.

6. The Renaissance (1400–1600)

Apart from the contributions to physics made by Leonardo da Vinci (1452–1519), the 15th century was not as brilliant as one might have expected considering the promising start that had been made in scientific investigation and in science teaching during the earlier centuries.

However, during the 16th century an invention took place that was to be of inestimable value in the diffusion of knowledge of physics – the discovery of printing. Many of the titles issued by the early printing presses were of 13th- and 14th-century works on science. There were several editions of Grosseteste's treatise on optics. Roger Bacon's works were printed in 1614. Witelo's *Perspective* was printed in 1535, in 1551 and again in 1572, whilst Peckham's *Perspectiva* went through nine editions between 1482 and 1627. Hero's *Pneumatica*, which remained the standard textbook on the subject till the work of Torricelli, Guericke, Boyle and Mariotte, appeared in three Latin editions between 1575 and 1700. His *Mechanica* was printed in 1580. The works of Archimedes appeared in 1544, and in 1558 a member of Melanchthon's circle published Peregrinus's treatise on the magnet.

The discovery of printing led to the writing, in the vernacular, of numerous popular compendia of knowledge. The numerous editions through which many of these compendia and encyclopaedia went indicates the thirst of the people of those times for knowledge. The best known of these was probably the 'Pearl Philosophic' *(Margarita Philosophica)* of Gregorius Reisch, which was first printed in 1503. The subjects of astronomy, natural philosophy, chemistry, optics, etc. are treated in this encyclopaedia, which was illustrated and intended as a textbook for young students.

The most important aspect of scientific progress in the 16th century was the growing acceptance of observation, experiment and rationality – the 'scientific' attitude – in the place of the authority of the ancients. In 1536 Peter Ramus (1515–72) wrote in his MA thesis at Paris University that 'all that Aristotle has said is false'; and Marcellus Palingenius Stellatus voiced the same sentiment in his *Zodiacus Vitae*.[9] This attitude is also discernible in the writings of Leonard Digges, [10] his brother Thomas Digges[12], both of Cambridge University,[11] and Dr John Dee, of St John's College, Cambridge, all writing in the 1570s.

The brilliant achievements of, among others, Galileo in dynamics and Gilbert in magnetism, were the fruits of putting into practice this 'modern' experimental scientific attitude.

By the 16th century considerable industrial and commercial expansion was taking place, and this resulted in a greater demand by the rising middle classes for a more utilitarian education biased towards science and mathematics, and the substitution of a more realistic approach to life for the aloofness of the cloister.

The 16th century is significant for the publication of several educational treatises that paved the way for a new presentation of studies not only in the university curricula but also in that of the schools, and which encouraged realism in education in distinction to scholastic formalism. The writings of Ramus, Francis Bacon, Sir Thomas Elyot, Rabelais, Vives and Melanchthon all catch a glimpse of the future reserved to science education, to a study of Nature by inductive speculation, to a study of things instead of the worship of words.

For the traditional *quadrivium*, Ramus would substitute mathematics, physics (including astronomy), metaphysics and ethics. The textbook to be used in physics was his own treatise, *Studies in Physics*, which, in spite of his criticism of Aristotle, was based on the latter's *Physics*.

In the *De Tradendis Disciplinis* (1531) Vives advocated the study of physics, even in the schools. But the subject still needed to be systematized and simplified before it lent itself to instruction of the young. By the end of the Renaissance the ground had been prepared for 'modern' phyiscs to take off. Looking back from the vantage point of the 20th century one might dismiss the early efforts and achievements as not amounting to much. But it was by 'standing on the shoulders' of such forerunners that Sir Isaac Newton was able to 'see ahead'.

Chapter Two

The 17th and 18th Centuries
Sir Isaac Newton (1642–1727)

In the year 1642 Galileo, one of the greatest physicists of any age, died. It was the year, too, when an even greater genius was born – Isaac Newton. Newton was born on Christmas Day, in the village of Woolsthorpe, in Lincolnshire. He grew up under the care of his mother and grandmother as his father died before he was born.

At the age of 12 he was sent to the Grammar School at Grantham. By now 'modern' physics had been born. The achievements of earlier physicists had been crowned by the brilliant work of Galileo and Torricelli in Italy, of Guericke in Germany, of Huygens and Snell in Holland, of Mariotte and Descartes in France, and of Boyle, Hooke and Halley in England. Physics rather than metaphysics, experimentation rather than disputation, had increased from a mere trickle into what was soon to become a flood; so much so that science was in danger of being reduced 'to a worship and idolisation of experiment as an end in itself'.[1]

The doors of the grammar schools, however, remained closed to the 'new' physics. It had taken generations to build up the classico-religious tradition and every attempt to obtain a lodgement of a new subject into these strongholds of conservatism and tradition was resisted. The curriculum consisted of the medieval *trivium* and *quadrivium*.

The new scientific curiosity of the 17th century did, of course, spawn demands for more practical and experimental teaching based on science. The writings of Hezekiah Woodward, Sir William Petty, John Dury and Samuel Hartlib called for a revolution in the curriculum and methods of schoolteaching: the monopoly of the classics must give way also to the study of science. 'Why should we not,' Comenius asked, 'instead of dead books, open to children the living book of nature?'[2]

In his *Great Didactic* (1630) Comenius lamented that 'hardly anyone teaches Physics by ocular demonstration and experiments, but all by recitation of the texts of Aristotle, or of some other.'[3] To remedy matters Comenius himself wrote his *Orbis Pictus* (1658) which was an illuminated Latin reader, 'a little encyclopaedia of things subject to the senses', in which many items of simple physics are mentioned.

In 1648 Sir William Petty argued for the substitution of the grammar school by a trade or industrial school where children would be 'taught ... Making Mathematicall Instruments ... Grinding of Glasses Dioptricall and Catoptricall ... Making Musical Instruments ... Mariners Compasses, Globes and other Magnetick devices.'[4]

Another pedagogue who played an important role in the diffusion of the new ideas concerning increased study of Nature and of a more realistic and experimental approach to such studies was John Dury (1596–1680). In his *Reformed School* (1650) he urged the study of science:

> all the Usefull Arts and Sciences...the Opticks with the Instruments belonging thereunto ... the Principles of Naturall Philosophie.'

Moreover

> The large common room ought to be furnished with all manner of Mathematicall, Naturall, Philosophicall and all other sort of pictures, maps, globes, instruments, models, engines that at the times appointed, the Ushers may leade their Scholars into it; to receive the lessons which they shall give them, upon the ocular inspection of the Things, which shall be shewed unto them.[5]

In 1642 Dury wrote letters proposing that books for simple science teaching should be specially written but, unfortunately, nothing seems to have come of the proposals.

George Snell, in his *On the Right Teaching of Useful Knowledge* (1649) proposed that optics and other branches of physics be introduced into the schools. Like Dury and Hartlib, he realized that before a reformation could be effected in the schools, schoolmasters had to be reformed. And to do this he proposed 'Colleges of Teachers, otherwise there will not be found in any place, a sufficient number of men abel to teach all the needful knowledge.'[6] In his *Thoughts Concerning Education* (1668) Bishop Gilbert Burnet FRS argued the case for experimental science in the schools.

Newton was born too early to benefit from any of these reforms. What he did learn, though, was sufficient Latin and English to be able to write one of the greatest physics textbooks ever written, the *Principia*, as easily in Latin as in English.

Teaching at Oxford and Cambridge in the 17th Century

In 1660, at the age of 18, Newton entered Trinity College, Cambridge, where his uncle had been a student. Many of the university students were no older than schoolboys of today. Though the average age of entry was about 16, many entered much earlier. Dr John Wilkins and Ralph Bathhurst, both founders of the Royal Society, entered when only 13 and 14 respectively. Christopher Wren went to Oxford when he was only 14 and made remarkable progress in astronomy, statics and mechanics.[7]

Aristotelianism persisted in the universities long after the modern scientific attitude had shown itself. At the disputations, as Whewell said, 'the same knots were tied and untied: the same clouds were formed and dissipated by book-learned scholars who frowned on studies that were practical'.[8] The 1431 Oxford Statutes had laid down for the BA degree two terms' work on one of the following books of Aristotle: *The Physica, De Coelo et Mundi, De Plantis, De Anima, De Animalibus.* In 1636 the Laudian Statutes prescribed that the Sedleian Lecturer in Natural Philosophy should expound these same works. This Revised Code, with a few additions, was to remain in force at Oxford down to the University Reform Act of 1854. The long list of Waynflete Readers of Natural Philosophy at Oxford, none of whom left any original work, shows how barren discourse on this subject must become when it is founded on Aristotle rather than on nature. The dominance of Aristotle in the universities was, however, increasingly questioned as a result of the advances made by Galileo and several other experimental physicists.

Not surprisingly numerous critics raised their voices against the prevailing methods and general neglect of experimental science teaching in the universities. John Webster (1610–82) not only criticized the defects of the existing curriculum in his *Academiarum Examen* (1653) but suggested that the balance, overwhelmingly in favour of classical studies, be redressed by including and giving preference to physics and mathematics, subjects of practical value. He pleads thus for the experimental method in science:

> O that the Schools therefore would leave their idle, and fruitless speculations, and not be too proud to put their hands to the coals and furnace ... nor be ashamed to cast off thy fine clothes to work in a laboratory ... that time may not be vainly spent in needless altercations, disputations, speculations, and observations ... that so they may not be sayers, but doers, not idle speculators, but painful operators ... that as nature hath given them two hands, and but one tongue, so that they may learn to work more and speak less.[9]

However, Dr John Wilkins, Warden of Wadham College, and Dr Seth Ward, Savilian Professor of Astronomy at Oxford, in their *Vindiciae Academiarum* (1654) show that the picture was not as black as it is usually painted. By this time the remarkable progress of mathematics made at Oxford under Ward and Willis and the progress made at Cambridge in the physical and mathematical studies, cut the feet from under many of these criticisms. Their statements show that Aristotelianism was beginning to be forced from its citadel, and that the supremacy of the old dialectical and philosophical studies was about to give place to experimental and mathematical disciplines. They write:

> And though we do very much honour Aristotle for his profound judgment and universal learning, yet we are so farre from being tyed up to his opinions, that persons of all conditions amongst us take liberty to dissent from his, and to declare against him according as any contrary evidence doth engage them, being ready to follow the Banner of Truth by whomsoever it shall be lifted up.[10]

Ward categorically denies that the mathematical and physical sciences were neglected. Concerning optics he says:

> ... there have been lately given here instances of more solid knowledge of all sorts of radiation or vision, than ever were here, and that such things are ordinary now amongst us (done by some amongst ourselves) as heretofore were counted Magicall ... I believe there is not one man here who hath not received the Copernican System ...
>
> Magneticall Philosophy is not neglected ... Instruments are prepared for those experiments (as Loadstone, a Terulla, and diverse others) and it is a reall designe amongst us, wanting only some assistance for execution, to erect a Magneticall, Mechanicall, or Opticke Schoole, furnished with the best Instruments, Adapted for the most usefull experiments in all these faculties ...

There is one thing which this sort of pamphleteers critics insist on. It is, that instead of verball Exercises, we should set upon experiments and observations, that we should lay aside our Disputations, Declamations and Publick Lectures ...

I have not neglected occasionally to tell the World, that this way is pursued amongst us.[11]

Experimental natural philosophy did not become a recognized academic study in the universities even in the second half of the 17th century, and much of the 18th century was occupied in marking time.

Whilst it must be conceded that, generally speaking, the heavy hand of the classical tradition persisted in the universities, an examination of the actual practices of the universities, of the colleges, and of the lives and interests of the university students and teachers, shows that the picture painted by historians, though dark, is by no means as black as they would have us believe.

By about 1650 the physics of Aristotle had slowly been replaced by that of Descartes (1596–1650) in the universities. Descartes' emphasis on the sufficiency of human reason marked the beginning of modern philosophy. In his *Discours de la Methode* (1637) he advocated a 'practical' philosophy in place of the 'speculative' which characterized the schools of learning. G. Compayrè summed up Descartes' educational influence:

Descartes not only reformed science and revolutionised philosophy, but banished from the schools the old routine, the mechanical processes and exercises of pure memory, and made a demand for rational methods that excite the intelligence, awaken clear and distinct ideas, and provoke judgment and reflection.[12]

By 1646 Cartesianism was exerting influence at Cambridge. Henry More (1614–87), fellow and tutor of Christ's College, was teaching the Cartesian physics. Descartes' physics became increasingly familiar at Cambridge and was studied in English and Latin translations of the *Traite de Physique* (1671) of the Cartesian Jacques Rohault (1620–75). This textbook dealt with matter, motion, reflection, refraction, vision, the elements, heat, astronomy, meteorology, geography and also physiology and medicine. Such Cartesian textbooks held the field till the first quarter of the 18th century when they were replaced by the Newtonian physics.

As an undergraduate Newton would have read Descartes' physics. He would also have studied the works of Kepler, especially his *Optics*. Newton also became acquainted with the *Optics* of Dr Isaac Barrow (1630–77). Barrow had entered Trinity College, Cambridge, when 14, and was appointed to the Chair of Greek at Cambridge in 1660. At this time, Newton, at Trinity College, had begun his experiments on the incidence of white light on a prism. Indicative of the shift in interest towards the new experimental philosophy was Barrow's lament, while Professor of Greek that 'I sit lonesome as an Attic owl, who has been thrust out of the companionship of all other birds; while classes in Natural Philosophy are full.'[13]

In 1662 Barrow was appointed to the Lucasian Chair of Mathematics, the only endowment in the physical sciences at the time. The professor was to lecture on geometry, arithmetic, astronomy, geography, optics, statics or other branches of mathematics. Newton attended Barrow's lectures on optics. In these *Lectiones*

Opticae, later published in 1669, Barrow dealt with reflection and refraction, with the properties of lenses, and with a simplified Cartesian explanation of the rainbow. It seems that Barrow's lectures failed to attract any considerable audience.[14]

At this time Newton himself was carrying out optical experiments on lunar haloes, on the problem of colour and on constructing a reflecting telescope. He was also working on his new method of fluxions and on analysis of infinite series – the 'germs' of the calculus. In 1665, aged 25, he was sent down from Cambridge because of an outbreak of plague, and returned to his home in Lincolnshire. While he was at home Newton worked out the fundamentals of the theory of gravitation, and devised instruments for grinding lenses. It is possible he also performed his classic experiment on the decomposition of light using glass prisms at this time.

On his return to Cambridge two years later Newton was elected a Fellow of Trinity College. It is very probable that he took pupils, as was the custom then. In 1668 Newton was awarded the MA. It was in that year that he made the first reflecting telescope, with a magnifying power of 40. Newton believed that it was the production of colours by refraction in lenses that had prevented improvement in the efficiency of telescopes. Newton's reflecting telescope excited considerable interest and it was even shown to King Charles II.

The Royal Society was founded in 1662. It grew out of the weekly 'experimental philosophical clubbe' or Oxford Philosophical Society, which had been established by Dr John Wilkins (1614–72), the Warden of Wadham College, Oxford. Wilkins, 'as well seen in Mechanicks and Experimental Philosophy as any Man in his time, and ... a great Promoter of them',[15] was a prominent figure who did much to stimulate the study of the physical sciences. In 1648 he published the first English textbook on mechanics, his *Mathematicall Magick ... concerning Mechanical Powers and Motions*. A chapter is devoted to each of the mechanical powers – the balance, lever, wheel, pulley, wedge and screw. Wilkins is familiar with the principle of moments, with the manner of determining the position of the resultant of a set of parallel forces, and he calculates the mechanical advantage of the machines he describes. This textbook of mechanics was reprinted in 1680 and again in 1708.

Another to give much encouragement to Oxford science was Christopher Wren (1623–1723), the Savilian Professor of Astronomy. In the 'Catalogue of New Theories, Inventions, Experiments and Mechanick Improvements, exhibited by Mr Wren, at the first assemblies of Wadham College in Oxford, for Advancement of Natural and Experimental knowledge' are listed:

The Weather-wheel, a Weather-cock,
The Libra Expansionis Aeris
The Ballance
Artificial Eye; a Perspective box
Divers new Engines for raising of water; Pneumatick Engines
Musical Instruments; a speaking Organ
Ways of submarine Navigation; a Compass.

John Wallis (1616–1703) was appointed Savilian Professor of Geometry at Oxford by Cromwell in 1649. In 1668 he laid down the principles of collision of imperfectly elastic bodies. He followed this in 1669 with a work on statics and in 1670 with a work on dynamics. His *Mechanica* issued during 1667–71 was one of the most comprehensive works on the subject then existing.

As regards the studies at Oxford, Wallis writes:

> I do not know any part of usefull knowledge proper for scholars to learn; but that if any number of persons desire therein to be informed, they may find those in the university who will be ready to instruct them: so that if there be any defect therein, it is for want of learners not of teachers.[16]

For the next few years Newton was mostly absorbed in chemistry and optics. He was also busy revising and editing Barrow's 'Lectures'. Barrow recognized Newton's genius and in 1669 was happy to resign in favour of his illustrious pupil, while he moved on to the older and more prestigious Chair in Divinity. Newton was then 27 years old.

When Newton was elected Lucasian Professor of Mathematics in October 1669 he chose optics and mathematics for his lectures and researches; and before the year had ended he had worked out the details of his discovery of the decomposition of light, and from this knowledge proceeded to give a complete explanation of the rainbow. Newton also investigated the colours of thin plates and coloured rings, the bending of light and the coloured fringes at narrow slits. His observations on double refraction in Iceland spar laid the foundations for the theory of the polarization of light. In the controversy over the theory of light Newton threw his great authority on the emission corpuscular theory, with the result that the wave theory of Hooke and Huygens was in abeyance for over a century. These discoveries formed the subject matter of the lectures he gave as Lucasian Professor during the years 1669 to 1671. In 1669 these were published as *Optical Lectures read in the Publick Schools of the University of Cambridge*. The manuscript of these lectures was reprinted in 1729 as *Lectures Opticae*. In these Newton deals with the decomposition of light by a prism, together with a detailed account of his famous experiment. In the second section Newton details the method he invented for determining the coefficients of refraction of different substances by making a ray pass through a prism of the material so that the angle of incidence is equal to the angle of emergence. Newton also deals with refraction both at plane and at curved surfaces, and applies his theory of colours to the rainbow.

Being of shy disposition, Newton does not seem to have been particularly interested in teaching or to have attracted many students to his courses, with the result that 'oftentimes, for want of learners' Newton suffered the experience of reading his lectures 'to the walls'.[17]

The *Principia*

Even more important was Newton's *Principia*, published in Latin in 1687. In this Newton defined mass, force, momentum, acceleration, etc. clearly for the first time and worked out his Laws of Motion. The rise of the Newtonian school of physics at Cambridge dates from about 1690. In 1695 Samuel Clarke (1675–1729), an undergraduate of Caius College, defended a question taken from Newton's natural philosophy for his bachelor's degree. In 1697 Clarke made a Latin translation of Rohault's *Physics*, but in his footnotes he included refutations of Descartes based on Newton's views. Through this strategem Newton's ideas were forced on the attention of the more conservative tutors. This Rohault-Clarke textbook, which went through several editions, became the outstanding physics textbook not only in England, but also in America. It was translated into English in 1723 by John Clarke, Samuel's brother.

The universities of St Andrews and of Edinburgh have the distinction of being among the first places where Newtonian physics was taught. Soon after the publications of Newton's *Principia* in 1687 Dr David Gregory (1661–1708), the Professor of Mathematics at Edinburgh, introduced Newton's physics in his lectures on optics, dynamics, mechanics and astronomy. In 1692 Gregory was elected to the Savilian Professorship of Astronomy at Oxford, where he gave regular systematic teaching in mechanics, optics, hydrostatics, astronomy, and the mathematical portions of music. His teaching consisted of 'explaining the proposition, and illustrating it with examples, operations, experiments, and observations, as the matter shall require, until the company apprehend and understand it.'[18] In 1695 Gregory published his *Catoptricae et Dioptricae Elementa*, for which there was 'a great demand'. Gregory's *Astronomiae et Geometricae Elementa*, published in 1702, was one of the first textbooks based on Newton's gravitational principles. James Gregory (1666–1742), brother to David, emulated his brother in his teaching of Newtonian physics in his capacity as Professor at St Andrews University.

Another to teach Newtonian physics early in the century was Gregory's favourite Edinburgh pupil, Dr John Keill FRS (1681–1721). In 1697 Keill, who was lecturer in natural philosophy at Hart Hall, Oxford, introduced the Newtonian philosophy in his lectures. Keill illustrated his lectures by experiments using apparatus he had devised himself. Keill's lectures were published in 1700 as *Introductio ad Veram Physicam*. The *Introductio* was welcomed as an excellent introduction to Newton's *Principia*, and went through several editions. The lectures covered mechanics (motion, momentum, centre of gravity, laws of motion, the pendulum, elasticity); hydrostatics (fluid pressure, Archimedes' Principle, specific gravity, siphon, pumps); pneumatics (air pressure, experiments with the air pump); barometer, thermometers, hygrometers; catoptrics (reflection, mirrors); dioptrics (refraction, convex and concave lenses, camera, 'magic lanthorn', the eye, defects of vision, microscopes, telescopes). Whilst the standard of this physics was probably about the same as that of the present day GCSE level it must be remembered that the undergraduate of those days was no older than most present-day GCSE candidates.

In 1710 that great teacher John Theophilus Desaguliers (1683–1749) of Christ Church gave instruction in experimental philosophy at Hart Hall in succession to Keill. He illustrated his lectures with 70 experiments on mechanics, 25 on hydrostatics, 25 on pneumatics and 30 on optics. In 1713 Desaguliers removed to London where he continued his course on experimental physics. Writing in 1745 he observes that the course on experimental philosophy he was then giving was the 121st since he began in 1710. His lectures were published in 1717 as *Physico-Mechanical Lectures ... wherein the Principles of Mechanics, Hydrostatics and Optics are Demonstrated and Explained by a Great Number of Experiments*.

About this time John Friend, of Christ Church, was giving chemistry lectures at the Ashmolean Museum and since 'heat' and even parts of 'magnetism and electricity' were then studied under 'chemistry', some knowledge of these branches of physics was given to his audience. The Ashmolean Museum, complete with laboratory, lecture room, library and exhibition gallery, was built in 1679 to remedy the lack of a centre for scientific studies in Oxford.

By 1716 lectures in experimental natural philosophy (embracing mechanics, optics, hydrostatics and pneumatics) were also being given in the Ashmolean Museum by J. Whiteside. A. Charlett wrote of Whiteside's course that

> He goes through all the parts of Natural Philosophie Experiments and Mathematics, for which he is very well accomplished, with excellent instruments well made, at expense of nearly £300. His price is a giny and a half.[19]

Whiteside divided the entrance hall of the Museum into two: in one half he would deliver his lecture; at the end the class would remove to the other half where an assistant exhibited the experiments. These lectures were continued until 1728.

James Bradley (1693–1762) who had succeeded Keill in 1729 and who had purchased Whiteside's apparatus – 'the best that is known' – for over £400, continued the courses on experimental physics in the Ashmolean until 1760 when he gave his 79th and presumably last course. His lectures covered mechanics, optics, hydrostatics, pneumatics, magnetism, electricity and astronomy. The average number that attended Bradley's courses is reported to have been 57. Several of those who attended these courses later became schoolmasters and parsons and, no doubt, helped in the diffusion of knowledge of experimental philosophy by their later activities.

The increased attention paid to experimental physics is evidenced by the fact that Oxford could boast a reputable vendor of scientific apparatus, namely Jean Prujean, who lived near New College; and in 1731 the Convocation of Oxford set apart £30 for a 'reader of experimental philosophy'.

Dr Thomas Hornsby (1733–1810), the Professor of Astronomy, continued Bradley's courses on experimental physics in the Ashmolean Museum. His lectures were well attended. In 1782 he was appointed to the Sedleian Chair of Experimental Philosophy. That Hornsby had at his disposal a vast collection of apparatus with which to illustrate his lectures is evidenced by the fact that in 1790 Edward Nairne valued it at £375 14s 6d. Edward Leigh, on his death in 1786, left £1000 in trust to Oriel College, Oxford, for models and apparatus to illustrate lectures on experimental physics. As a result an imposing collection of physics apparatus came into the possession of the college.

Teaching at Oxford and Cambridge in the 18th Century

The 18th century saw many new discoveries which were passed on to students through lectures at universities and elsewhere. This was the century of Dr Joseph Black's studies in heat, James Watt's improvement of the steam engine, Stephen Gray's electrical experiments in Corpus Christi College in 1708, and

Musschenbroek's discovery of the Leyden jar (1740). By about 1745 electrical experimentation had become so popular that numbers of persons gained their livelihood by giving public demonstrations of electrical experiments. In 1752 Benjamin Franklin demonstrated that lightning was an electrical phenomenon; in 1789 Galvani discovered current electricity; and in 1800 Volta described the voltaic pile.

Some idea of the physics teaching in the universities early in the 18th century is provided by the study of schemes recommended for undergraduates. Daniel Waterland (1683–1740), a tutor at Magdalen College, Oxford, in his *Advice to a Young Student, with a Method of Study for the Four First Years* (1706) recommended for the second year Wells's *Astronomy*, Whiston's *Astronomy*, Keill's *Introduction to the True Physics*, Rohault's *Physics*, Bartholini's *Physics*, Wells's *Mechanics, Statics and Optics* and Le Clerc's *Physics*; and for the fourth year Newton's *Optics*, Whiston's *Physical Mathematics* and Gregory's *Astronomy*. Other textbooks suggested are Newton's *Principia*, Ozanam's *Cursus Mathematicus*, Molyneaux's *Dioptrics*, Harris's *Lexicon*, and the works of Sturmius and Huygens.

Robert Green, Fellow and tutor of Clare College, Cambridge, in his 'Scheme of Study' (1707) recommended for the second year Descartes, Rohault, Le Clerc and Boyle; for the third year experimental philosophy, including the advances of Boyle and those recorded in the *Philosophical Transaction* and in the *Collegium Curiosum*; for optics Newton, Descartes, Rohault, Kepler, Molyneux, Gregory, Dechales, Huygens and Barrow; and for the fourth year mechanical philosophy, statics, hydrostatics, gravitation, the authors being Marriot, Keill, Huygens, Sturmius, Boyle, Newton, Ditton and Wallis.

That the Newtonian physics was being taught at Cambridge early in the century is evidenced by William Stukeley (1687–1765), who went to Corpus Christi College in 1703, and who states that 'We had then in our college, under the instruction of Dr Robert Dannye ... gone thro' an excellent course of lectures in mathematics, and philosophy, particularly the Newtonian.'[20] Furthermore, he saw 'many Philosophical Experiments in Pneumatic Hydrostatick Engines and instruments ... and the doctrine of Optics and Telescopes and Microscopes, and some Chymical experiments.'

By 1707 public exercises in the Schools, founded on every part of the Newtonian philosophy, had become very common. Richard Laughton (1668?–1723), who lectured on the Newtonian physics at Clare Hall, published a paper of questions on this new physics, suggesting them as suitable theses for the disputations in the Sophist Schools.

The popularity of the Newtonian physics is instanced by Sir William Browne's statement in 1772 that: 'Sir Isaac Newton's Mathematical Principles of Natural Philosophy or Knowledge, had risen so high above par, that I gave no less than two guineas for one, which was then esteemed a very cheap purchase.'[21]

The magic of Newton's name ensured that the physical sciences were anything but neglected at Cambridge. In 1701 William Whiston (1667–1752) succeeded Newton in the Lucasian Chair. Whiston's lectures on the Newtonian physics were published as *Praelectiones Astronomiae* in 1707 and in 1710 as *Praelectiones Physico-Mathematicae*. In 1710 Whiston was expelled from Cambridge on religious grounds. He removed to London where, in collaboration with Francis Hauksbee, he gave a series of lectures on physics, illustrated by demonstrations and experiments. The course covered mechanics, optics, hydrostatics and pneumatics.

Whiston was succeeded by the celebrated blind mathematician Nicholas Saunderson (1682–1739) who gave lectures on the Newtonian philosophy, hydrostatics, mechanics, sound, astronomy, tides and optics. Saunderson, who 'though he had not the use of his own eyes, taught others to use theirs', drew large audiences.[22]

In 1704 the Plumian Chair of Astronomy and Natural Philosophy was established. Roger Cotes (1682–1716), the first holder of the Chair, was another pioneer of experimental methods and in 1707, along with Whiston, started a course of lectures on hydrostatics and pneumatics in Trinity College, a 'Course of Experiments, [being] actually performed and explained before the reading of each Lecture'. In 1738 Cotes's professorial lectures were published by his successor in the Plumian Chair, his cousin Robert Smith, as *Hydrostatical and Pneumatical Lectures*. These were issued for a third time in 1775.

Robert Smith (1689–1768) lectured as Plumian Professor on optics and hydrostatics. In 1738 appeared his *A Compleat System of Opticks* which had a wide circulation and in which he proved his assertions by experiments. This textbook was reprinted in 1778. In 1749 Smith published his *Harmonics*, a textbook of sound, which contained the substance of his lectures. Smith lectured as Plumian Professor till 1760.

During 1743–8 Thomas Rutherforth (1712–71) lectured so successfully on natural philosophy at St John's College that he was appointed a Regius Professor of Divinity as a result. The English edition of his lectures appeared in 1748 as *A System of Natural Philosophy, being a course of Lectures in Mechanics, Optics, Hydrostatics and Astronomy*. In the dedication, Rutherforth remarks that the 'new' natural philosophy 'was looked upon as a necessary branch of academical learning'. His lectures, which covered mechanics, optics, hydrostatics, pneumatics, sound and astronomy, can be compared with today's GCSE in physics.

As a result of the efforts of men such as those listed above it was possible for John Green to write in *The Academic: or, a Disputation on the University of Cambridge* that:

> Mathematics and Natural Philosophy are so generally, and so exactly understood, that more than 20 in every year, of the Candidates for a Bachelor of Arts Degree, are able to demonstrate the principal Propositions in the *Principia*; and most other Books of the first Character on those Subjects ... And others, are yet well acquainted with the Experiments and Appearances in Natural Science.[23]

Anthony Shepherd (1712–96) succeeded Robert Smith in 1760 as Plumian Professor of Astronomy and Experimental Philosophy. In about 1770 he published his *Heads of a Course of Lectures in Experimental Philosophy, read at Christ College*. These covered mechanics, electricity, magnetism, astronomy, hydrostatics, pneumatics, and optics. However, Shepherd evidently soon wearied of lecturing and in 1776 appointed two fellows, Brand and Metcalfe, as deputies to read lectures for him. These two lectured very successfully on physics, as is attested by Philip Yorke, who, writing in May 1776, states:

> I have been attending the experimental lectures this last week and have been much entertained with them. Mr Brand, who gives the mechanical part, has shown the experiments with very good success, for except one or two of the

magnetical ones, scarcely any have failed. I have been extremely entertained with the experiments on hydrostatics and particularly with the invention and improvements of the air pump.[24]

About the same time George Atwood (1746–1807), fellow and tutor of Trinity College, was giving a course of experimental lectures in physics open to all students. In his Trinity lectures in 1782 and 1783 he used the machine which is still known by his name for finding the acceleration of falling bodies. In 1776 he published a description of the experiments used in his lectures. Fifty experiments were used in his mechanics, 42 in hydrostatics, 33 in electricity, 10 in magnetism and 54 in optics. Many of these experiments were similar to those used in modern GCSE physics courses.

In 1783 the Rev. Richard Jackson founded a Chair of Natural and Experimental Philosophy. Sir Isaac Milner (1750–1820) was the first Professor. He gave alternate lectures in chemistry and in experimental philosophy. These lectures covered matter and motion, heat, fluids, pneumatics, chemistry, electricity, magnetism and optics. Whilst the general standard of his courses compares with the modern GCSE physics, in some instances it approaches Advanced level. Milner took great pains with his lectures, and Professor William Smyth thought him 'a very capital lecturer', adding that 'what with him and his German assistant Hoffman, the audience was always in a high state of interest and entertainment'.[25]

Milner was succeeded in 1792 by Francis J.H. Wollaston (1762–1828) who certainly fulfilled the conditions of his Chair, giving lectures alternately on chemistry and experimental philosophy, illustrated with not less than 300 experiments annually.

In 1796 Samuel Vince (1749–1821) succeeded Shepherd in the Plumian Chair, but three years before this he was giving regular courses in experimental philosophy, including Newton's *Principia*. Vince's *Plan of a Course of Lectures on the Principles of Natural Philosophy* was published in 1793. The Lectures covered the usual topics of mechanics, hydrostatics, optics, magnetism, electricity and astronomy. In the Plan Vince gives copious notes along with details of experimental proofs and demonstratations. In 1795, in collaboration with the Rev. James Wood, Vince published *The Principles of Mathematics and Natural Philosophy*, which became a standard university textbook. Vince held the Plumian Chair till his death in 1821. In 1796 Wood (1760–1839) published *The Principles of Mechanics*, followed two years later by *Elements of Optics*. Both these textbooks formed part of the series known as the 'Cambridge Course of Mathematics'.

Existing advertisements show that in the 1790s regular courses in physics were being given in the University and in some colleges, quite apart from the statutory lectures of the Professors. An advertisement dated Trinity Hall, No. 2, 1793, draws attention to a course on natural philosophy, including Newton's *Principia*, to be given by the Rev. F. Wrangham (1769–1804). The advertisement of his 1802 lectures reads as follows:

> The lectures are experimental, comprising mechanics, hydrostatics, optics, astronomy, magnetism and electricity, and are adapted to the plan usually followed by the tutors in the university. All the fundamental propositions in the

first four branches are proved by experiments, and accompanied with such explanations as may be useful to the theoretical student. Various machines and philosophical instruments are exhibited in the course of lectures and their construction and use explained. And in the two latter branches a set of experiments are instituted to shew all the various phenomena, and such as tend to illustrate the different theories which have been invented to account for them ... Terms are three guineas for the first course, two guineas for the second, and afterwards gratis.[26]

During the second half of the century an important change was made in the method of examinations at Cambridge. Up to that time degrees had been awarded on disputation. With the introduction of the Senate House Examination all candidates for the BA, in addition to disputation had to take a *viva voce* examination and dictated questions. (Printed question papers were sparingly used before 1836 when the whole machinery of examinations was recast.)

This new examination was to have a very beneficial effect regarding physics teaching since it made the study of mathematics and of certain branches of physics compulsory for all candidates. Questions were taken from mechanics, hydrostatics, astronomy and optics as explained in the works of Cotes, Helsham, Rutherforth, Keill, Long, McLaurin, Smith and Newton. The Rev. John Jebb deprecated the over-emphasis given the mathematics and physics in this examination. He complains:

But as the highest Academical Distinctions are invariably given to the best proficients in Mathematics and Natural Philosophy, a very superficial knowledge in Morality and Metaphysics will suffice ... It is only to be lamented, that merit in Mathematics and Philosophy should be so rewarded, so much to the exclusion of every other kind of Literary Improvement.[27]

In order to obviate the tendency towards specialization in the more abstruse parts of mathematics, the Vice-Chancellor, in a Declaration dated 20 May 1774, stipulated that no attention would be paid to a candidate's achievements in higher mathematics unless he showed a thorough acquaintance with mechanics, hydrostatics, optics and astronomy. This Declaration thus gave still greater encouragement to the study of physics. Success in the new examinations was ensured by the intensive study of 'Syllabuses' of physics, later called 'College manuscripts', which were published in considerable number to meet the demand which had arisen for them.

An examination of the physics teaching in the universities of both Oxford and Cambridge at this time reveals that they were not as moribund as general opinion imagines. Dr N. Hans concluded that about 80 per cent of the 380 English scientists of the 18th century were educated at Oxford and Cambridge; and of these 180 specialized in natural philosophy, 122 in mathematics and astronomy, and 34 in chemistry.[28]

Evidence of the diligence of the university tutors is provided in a pamphlet, *Observations on the Present State of the English Universities* (1759). The author asserts that:

there never were more able tutors, nor ever properly and worthily employed, than at present ... never was a larger portion of true useful science taught, nor in so good a method ... This has been evidently the case of late in most, if not all, the chief branches of philosophy, both natural and moral ... And though, with respect to the former of these, which is so extensive a nature, and where so many instruments and experiments are requisite, that a suitable apparatus might seem above the purse of most private fellows of colleges, there must be very considerable difficulties; yet are these in a good measure conquered, and a great advance has been made in it, in truth much greater, than could well be expected from the small pittance commonly allowed to tutors.[29]

Vicesimus Knox, late fellow of St John's College, Oxford, in his *Liberal Education* (1781) also paid a striking compliment to university teaching of experimental physics. He informs us that:

> The lectures in the universities on these subjects [astronomy, and natural and experimental philosophy in all its branches], are admirably well fitted to accomplish the ingenious pupil in these delightful and improving sciences. In these places, a large and costly apparatus is always at hand, and the professors, who read lectures, are for the most part men of great and solid merit, with little ostentation.[30]

Christopher Wordsworth likewise spoke very highly of the Cambridge lecturers and tutors in the mathematical and physical sciences. In his *Scholae Academicae* (1877 edition) he says:

> ... if they had made enquiry in term time they would have found Roger Cotes of Trinity, Daniel Waterland of Magdalene, Nicholas Saunderson of Christ's, Christopher Anstey and J. Newcombe of St. John's, and I know not who beside, with well-filled lecture rooms in 1710. And if they failed to find in his college auditorium their friend Richard Laughton of Clare, the popular 'pupil-monger', it would be only because then, as in the preceding year, he was proctor, and ... fulfilling the office of moderator in the schools for the University at large, where he was encouraging the senior sophists and questionists to adopt the Newtonian philosophy in the exercises for their bachelor's degree.[31]

The Scottish Universities

Towards the end of the 17th century the shaking of the dry bones of scholastic medievalism had begun in the Scottish universities, and by about 1730 Newton was beginning to reign supreme.

Edinburgh

Robert Stewart, the Professor of Natural Philosophy (1709), having been brought up in the Cartesian system, at first taught this philosophy, but during the latter part of his professorship he taught the Newtonian physics.

The *Scots Magazine* of August 1741 (pp.371–2) gave the following account of Stewart's courses:

> He teaches, first, Dr John Keill's *Introducio ad veram Physicam*, and the *Mechanics* from several other authors. After that he teaches Hydrostatics and Pneumatics from a manuscript of his own writing. Then he teaches Mr. David Gregory's *Optics*, with Sir Isaac Newton *Of colours*; describes the several parts of the Eye, and their uses, with the Phaenomena of Vision; and describes all the different kinds of microscopes and telescopes. Then he teaches Astronomy from Mr. David Gregory's *Astronomy*, with some propositions of Sir Isaac Newton's *Principia*, and the Astronomical observations both ancient and modern. He likewise shews a set of Experiments, Mechanical, Hydrostatical, Pneumatical and Optical.[32]

In 1742 John Stewart was appointed conjunct Professor of Natural Philosophy with his father. John was succeeded in 1759 by Adam Ferguson (1723–1816), who drew crowded classes to his lectures. Mr John Robison, who succeeded to the Chair in 1774, gave courses covering mechanics, hydrostatics, hydraulics, pneumatics, astronomy, optics, electricity and magnetism and also the history of science. In 1777 the Town Council allowed him £300 for the purchase of physics apparatus. The grant of such a large sum of money for science apparatus emphasizes how responsible authorities had come to recognize the necessity of experimental work.

During the 18th century there was considerable overlapping as to what was taught by the Professors of Natural Philosophy and of Mathematics. Thus whilst Robert Stewart was teaching natural philosophy this subject was also being taught by the Professor of Mathematics, James Gregory, and when Gregory resigned, by his successor Colin McLaurin (1698–1746). In addition to mathematics, McLaurin lectured on optics, astronomy, Newton's *Principia* and a 'college of experimental Philosophy.' In 1748 McLaurin published *An Account of Sir Isaac Newton's Philosophical Discoveries*, which even today is one of the finest expositions of the Newtonian Philosophy in the English language.

Glasgow

By about 1704 some physics instruments and apparatus had been procured for the use of the regent professor in charge of the natural philosophy class. In 1712 funds were subscribed to augment this stock of apparatus. In 1727 Robert Dick was appointed Professor of Natural Philosophy. He conducted a course in 'experimental philosophy' twice a week. By 1730 Dick needed an assistant, Henry Drew, to help him in demonstrating the experiments. In 1752 Dick was succeeded by his son, who continued courses in mechanics, hydrostatics, pneumatics, optics and astronomy. 'In addition during April he carries on an experimental course at 4 p.m.'[33] During the period of about 30 years in which the two Dicks held the Chair over £300 was spent on physics apparatus. A complete apparatus for electrical experiments was purchased in 1749.

John Anderson (1726–96) formerly professor of oriental languages, succeeded the younger Dick in 1757. He very soon attracted an exceptionally large following of students. He continued the evening classes in experimental philosophy which had been begun by his predecessor. Though the students did not perform experiments themselves they were taken from the lecture room to the 'experiment room' to see the experiments performed. In 1768 Anderson complained that since his class in experimental philosophy had grown so large (from 150 to 200 students) he had had to divide it in two and deliver the same lecture twice in the same day, and also that the accommodation for demonstrating his experiments and for storing his apparatus was quite inadequate. The next year four rooms were converted into a lecture theatre, and a large apparatus room and a demonstration room were also provided. For the benefit of his students Anderson wrote *A Compendium of Experimental Philosophy* in 1760. This was followed by his *Institute of Physics* in 1786 which ran through five editions in ten years. This textbook dealt with the topics of mechanics, electricity, magnetism, gravitation, hydrostatics, hydraulics, pneumatics, sound, optics and astronomy. When Anderson died in 1796 he bequeathed his property to the establishment of a university giving scientific instruction and to be open to all classes and to both sexes. The Anderson Institution which was founded as a result boasts a most impressive record of able lecturers and brilliant students.

Thomas Garnett (1766–1802) was the first Professor of Natural Philosophy in Anderson's Institution. Large numbers of men and women attended Garnett's physics courses which were illustrated by experiments using 'apparatus [which] is the most extensive of any in this Island'.[34] Garnett continued his courses until 1799 when he was appointed first Professor in the Royal Institution, London.

Other colleges and universities

There was an experimental natural philosophy school at Marischal College, Aberdeen early in the century. In 1755 the curriculum of the College included natural and experimental philosophy, embracing mechanics, hydrostatics, pneumatics, optics, astronomy, magnetism and electricity. Patrick Copland (1749–1822), appointed Professor of Natural Philosophy in 1775, extended knowledge of physics beyond academic circles by means of his popular evening courses which he began in 1785.

At King's Colleges, Aberdeen, and at St Andrews University the curriculum also included experimental natural philosophy.

The Scientific Societies and the Mathematical Schools

Propagation of knowledge of the 'new' physics was also effected outside the portals of the universities. The chief educative influences in spreading this knowledge were the many scientific societies that sprang up around the country. The most prestigious of these was the Royal Society. The Royal Society received its Charter of Incorporation in 1662. It arose out of the informal gatherings of men drawn together through a common love of science, such as the meetings of 'virtuosi' who met at Gresham College. It was in order to cater for the needs of a society growing increasingly more conscious of the vital part that science could play in technology that Gresham College had been founded in London in 1596 for gratuitous instruction in the seven liberal arts and sciences. Gresham College was a place for lecturing to the general public and not a school for teaching boys or classes. Though no science instruction as such was given at the College, popular scientific lectures were given from time to time, as stipulated by the founder of the College.

About 1645, when the civil war interrupted academic studies at Oxford and Cambridge, men such as Dr John Wilkins, Warden of Wadham College, Dr John Wallis, the most distinguished mathematician before Newton, whose pupils included Christopher Wren, Robert Hooke and Edmund Halley, Dr Jonathan Goddard, Robert Boyle, William Petty, and John Evelyn, the diarist, met in London to 'inquire into natural philosophy'. On the purgation of the universities by Cromwell in 1648/9 many of these 'virtuosi' returned to Oxford and formed a branch of the 'Invisible College' – as Boyle called it – at Wadham College, where they held weekly an 'experimentall, philosophicall clubbe' and brought these studies into fashion there.[35] Christopher Wren later attended these meetings.

With tranquility restored by the restoration of the monarchy in 1660 members of this 'experimentall philosophicall clubbe' returned to London. At a meeting held at Gresham College a set of rules was drawn up for establishing a college for promoting 'Physico-Mathematical Experimental Learning'. A petition for incorporation was sent to King Charles II and on 15 July 1662, the Royal Society was set up, taking as its motto: *'Nullius in Verba'*. In his Diary of 15 February, Pepys, wrote: 'it is a most acceptable thing to hear their discourses and see their experiments'. In 1665 the *Philosophical Transactions* appeared for the first time and was soon recognized as the most important scientific periodical of the day.

King Charles II not only had a hand in the foundation of the Royal Society, he was also instrumental in the establishment of the Mathematical School at Christ's Hospital in 1673. The advice was sought of Wallis and Gregory at Oxford and of Newton at Cambridge for a syllabus for the School. Newton understandably pressed the claims of the 'science of Mechanicks'. He added that boys should also become fluent in Latin because 'the best Mathematical books are in that language'.

The Royal Society was not alone in fostering study of the new experimental science. Several Philosophical Societies (such as those of Edinburgh, Oxford, and Cambridge) sprang up all over England and even the Society at little Spalding could boast not only the patronage of Newton but also an air-pump, a thermometer and a barometer.

Scientific instruments were now available on a large scale for the first time. Among these may be mentioned barometers, thermometers, telescopes, microscopes, the camera obscura, lenses and prisms. A new profession – that of mathematical and scientific instrument-maker – arose to supply the demands of the new experimental sciences.

The Mathematical Schools

The growth of commerce and industry had resulted in the rise of a great merchant middle class who also clamoured for a more practical and utilitarian education for their sons. Several mathematical schools, which included at least some branches of physics in their curricula, were established to supply this demand. One such was the Christ's Hospital Mathematical School, founded in 1673. Newton, who was a governor of the school, had occasion to deprecate the lack of teaching in mechanics.

This was remedied in 1696 when the subject was introduced into the curriculum; the topics covered included force, motion, stresses, levers, pulleys, screws, anchors, pumps, rudders, guns and sails.

Physics in adult education

In 1706 a group of young men complained to the Lord Mayor of London that the wishes of the late Sir Thomas Gresham, as stated in his will, were not being carried out meticulously. As as result the Gresham Lectures were resumed. Lyson's *Collectanea*, a collection of advertisements from various newspapers, contains innumerable references to public courses and lectures given to cater for the growing demand for scientific knowledge, although some of these so-called courses were more in the nature of 'shows'.

Lectures in experimental physics were particularly popular. As early as 1704 Francis Hauksbee FRS (the elder, died 1714) was showing 'electrical, hydrostatical and pneumatical experiments' in London. In 1705 James Hodgson FRS was giving courses in experimental physics, illustrated 'with all sorts of Instruments; as engines for rarefying and condensing Air; ... with utensils for making Hydrostatical Experiments, Telescopes ... Microscopes ... with Prisms of several sizes for illustrating the New Doctrine of Light and Colours.'[36]

The fee for the course was two guineas. John Desaguliers FRS brought his lectures on experimental physics from Hart Hall, Oxford, to London in 1713. In 1727 they were illustrated by 70 experiments in mechanics, 25 in pneumatics and 30 in optics, and Desaguliers was assisted by an instrument maker, William Vream. The complete edition of Desaguliers' *Course of Experimental Philosophy*, published in 1734, came to be recognized as a standard work throughout the century, and his courses became a model to other lecturers on experimental philosophy.

These included William Whiston (1667–1752), formerly Professor of Mathematics at Cambridge (he had been removed for heresy), Francis Hauksbee the younger (1687–1763), and Dr Triboudet Demainbray (1710–82) who was appointed tutor in mathematics, experimental philosophy and natural history to the Prince of Wales (later George III) in 1754. Benjamin Worster, Thomas Watts and James Stirling held courses at the Little Tower Street Academy, James Dinwiddie gave lectures at the Lecture Room, Portugal Street and Martin Clare lectured at the Soho Square Academy. In 1716 Watts translated Rohault's *Treatise on Mechanics* into English for Humphrey Ditton, who wanted a textbook for his Mathematical School at Christ's Hospital.

Lectures were not confined only to London. Peter Shaw read his 'Chemical Lectures' in London in 1731–2 and later read them in Scarborough. The instrument-maker Benjamin Martin (1704–82), an ardent champion of the Newtonian philosophy and country-wide lecturer, taught mechanics and the construction and use of mathematical instruments at his school in Chichester, where pupils were encouraged to carry out practical scientific experiments as well as seeing these demonstrated. Martin also published several treatises on various aspects of experimental physics.

It was not only in London that courses in physics were given to adult audiences. Courses in mathematics and experimental philosophy were given in industrial and shipping centres such as Newcastle, for example. In 1711 James Jurin (1684–1760), an old boy of Christ's Hospital and a Cambridge graduate who had been appointed Headmaster of Newcastle Grammar School two years earlier, ran courses for adults on algebra and mechanics. The course was advertised as being 'greatly useful to all sorts of Persons, but especially to Gentlemen concerned in Collieries and Lead Mines'. It was to be given to a class of 10 or 12 three times a week for 12 or 18 months, the fee being an initial two guineas (£2.10) followed by a charge of half a guinea (52½p) per month.

The syllabus read

1. So much of the Principles of Geometry, Arithmetic and Algebra as shall be necessary for this undertaking.
2. The general laws of Motions, and the Principles of Mechanics deduced from them.
3. The Doctrine of Percussion, or the Effects which follow from the Stroke of Bodies upon one another.
4. The Natural Motion of all heavy Bodies.
5. The Motion of Bodies upon inclined Planes.
6. The Theory of all Kinds of Engines simple and compound, with a particular Explication of the Engines used in Collieries, and the method of Examining their Advantages or Defects.
7. Hydrostatics, under which Head will be demonstrated by Experiment: the chief Properties of Water and other fluids ...; The Method of Calculating the Weight or Pressure of Water against the Bank of Rivers, or Milldams ... and other Surfaces, and consequently determining the strength requisite for those Bodies to support the Pressure ...
8. Pneumatics, the Weight and Spring of the Air, its Refraction and Condensation ... the Air-pump, the Condenser, together with the Barometer, Thermometer, Hygrometer ... their Nature and Uses explained ...
9. Hydraulics, or the Doctrine of Water and other Fluids in Motion. The Method of estimating the Swiftness of Water running in any Canals open or closed, as in Rivers, or Mill-Races, ... Conduit Pipes etc. with the Quantities of Water that they discharge. A particular Application to the draining of Collieries ... for clearing and keeping a Colliery clear of Water. Of the Force of Fluids, as Air, or Water to carry about the Sails or Wheels of Mills and other Engines, and the best proportion of the Machines driven by them, or by Horses.
10. Lastly, the Important Theory of the Friction of Rubbing Machines, with the Impediment caused by the stiffness of the Ropes, for want of which the greatest Engineers have been disappointed in their undertaking, and the best concerted Machines have been rendered useless ... it will be explained in an easy manner, partly by Experiment, and the Application of it to the Calculation of Machines will be demonstrated.

For the accommodation of such gentlemen and ladies as do not choose to appear at the public school, I propose (at vacant hours) to attend them in their own apartments.

Newcastle also had the privilege of courses in mechanics, dynamics, hydrostatics, pneumatics, optics and astronomy, given by Isaac Thomson and Robert Harrison; and in the mathematical sciences by Charles Hutton, a local pit boy who became one of the leading mathematicians of his age, from 1760 to 1763.

Several towns throughout Britain benefited from the courses in experimental physics that were given by a tireless band of peripatetic lecturers who travelled from town to town. These included Caleb Rotherham, who lectured in Manchester and Newcastle, Samuel Kaye who gave a course of 24 lectures on physics in Manchester in 1743, James Arden, Benjamin Donnie and the eminent scientist James Ferguson (1710–76). Ferguson's courses, delivered in London and in the provinces (including Cambridge), covered the whole range of experimental physics. Much of the apparatus he needed he invented, designed and constructed himself. In addition to being a lecturer of renown on natural and experimental philosophy, Ferguson was also the celebrated author of a great many popular textbooks on physics. In 1760 his *Lectures on Select Subjects in Mechanics, Hydrostatics, Hydraulics, Pneumatics and Optics* was published.

The activities of Adam Walker (1731–1821) as a peripatetic lecturer in physics extended over 40 years. During this time he travelled through the north of England, the south of Scotland and even Ireland, giving lectures, illustrated by experiments, on mechanics, hydrostatics, pneumatics, chemistry, optics, astronomy, magnetism, electricity and general properties of matter. Such was Walker's reputation that he was engaged to lecture on natural philosophy in several public schools such as Eton, Westminster, Winchester, St Paul's and Rugby. During the 19th century Walker's youngest son, Deane Franklin (1778–1865)) continued his father's lectures at Eton, Harrow and Rugby. Apart from his numerous lecture courses Adam Walker also furthered the diffusion of knowledge of physics by the several textbooks he wrote.

Instrument-makers

Another group of men who rendered signal service in diffusion of knowledge of physics, were the makers of scientific instruments. Not only would further research in physics have been slowed down, but the teaching of physics using the experimental technique would have been impossible without the instruments and apparatus they made.

During the century several instrument and apparatus makers, such as the Hauksbees, Vream, Benjamin Martin and James Ferguson, set up business in London and the big towns to cater for the growing demand for scientific apparatus. Some of these men were also lecturers. George Adams the younger (1750–95), who succeeded his father as royal instrument-maker, was also the author of a large number of elementary physics books. In his *Essay on Electricity* (1784) Adams covered in great detail all that was then known of electricity and magnetism, mentioning no less than 226 experiments with full instructions on how they were to be carried out. In 1794 appeared his voluminous and well-illustrated *Lectures on Natural and Experimental Philosophy.*

Physics in Women's Education

By this time, a few women had already raised their voices for better education for women and girls. In 1641 Anna Maria von Schurman argued that women's studies could also include physics. In England Mrs Bathsua Makin, who was associated with members of the Royal Society, in 1673 advocated for women with leisure 'the whole Encyclopaedia of Learning', including experimental philosophy.[37] She kept a school for gentlewomen at Tottenham High Cross where 'those that think one Language enough for a woman, may forbear the Languages and learn only Experimental Philosophy'. Lady Mary Chudleigh (1656–1710) also recommended physics for women's education. Margaret Cavendish (1624–74), Duchess of Newcastle, in 1655 published her *Philosophical and Physical Opinions*, which was reprinted as *Grounds of Natural Philosophy* in 1668. In the same year appeared her *Observations on Experimental Philosophy*, dedicated to Cambridge University.

However, in spite of the example of these worthy women, the education of the 17th-century girl followed the three traditions of domestic arts, accomplishments and piety.

Chapter Three

Teaching in the Dissenting and Private Academies Joseph Priestley (1733–1804)

Joseph Priestley is renowned for his researches and discoveries in both physics and chemistry. On the basis of his electrical experiments he was elected to the Royal Society in 1766. Beginning in 1765 Priestley would spend a month each year in London where he could meet the leading men of science. Benjamin Franklin, for example, gave much encouragement to Priestley in his scientific researches. It was with Franklin's encouragement and generous loan of the requisite books that in 1766 Priestley wrote his *History and Present State of Electricity*, an original work in which he summarized the knowledge of electricity up to his time and described his own experiments. Priestley also carried out many original experiments with gases which led to the foundations of modern chemistry. Apart from his work in science Priestley's textbooks, his method of teaching and classroom demonstrations in science were major innovations in education.

Priestley was born of middle-class Calvinist parents near Leeds in 1733. When he was 12 he was sent to the local grammar school at Batley, where he began the study of Latin and Greek. The grammar schools were largely hide-bound by tradition and foundation statutes to a purely classical curriculum. Needless to say, there were many criticisms of this state of affairs.

Several 18th-century pedagogues were not slow to point out the defectiveness of an education restricted largely to two dead languages. More than that they pressed the claims for teaching science in the schools. In his *Essay upon Study* (1731) John Clarke (1687–1734), headmaster of the grammar school in Hull, advocated mathematics and natural philosophy, and recommended as the textbooks in physics Keill's *Introductio ad veram Physicam*, Gravesande's *Physices Elementa Mathematica Experimentis Confirmata* and Rohault's *Physics* supplemented by Dr Clarke's Notes. George Turnbull likewise advocated physics in the schools. In his *Observations upon Liberal Education* (1742) he pointed out that:

> Experiments to prove these (laws of motion ... gravity and springiness of the air ... specific gravities ... mechanical advantage) and many other such physical truths are easily performed, and as easily understood.[1]

James Burgh would also introduce schoolchildren to simple physics. Writing in his *Thoughts on Education* he states:

> I do not think it difficult to give most young people, who have reached the age of 12, though they be not intended for an academical education, a tolerable notion of theoretical astronomy ... A few experiments in natural philosophy such as those of the air pump, etc. ...[2]

The author of the anonymous tract *Education of Children and Young Students in all its branches* (1752) recommends that children ought 'to go through a course of experimental philosophy; a sight of the experiments will enable them to understand what they read.'[3]

At the same time as some progressive pedagogues were advocating the physical sciences in the education of children, others were actually writing simple textbooks adapted to the capacities of children. In 1761 Mr. Newberry published his *Newtonian System of Philosophy, adapted to the capacities of Young Gentlemen and Ladies.* This children's textbook covered the following topics:

Lecture 1. Of Matter and Motion – the Laws of Motion; Attraction and Gravitation; Magnet.
Lecture 2. Of the Universe – Eclipses; Velocity of light; Earth's motion.
Lecture 3. Of the Air – Atmosphere and Meteors; Elasticity of air; Wind gun; Light and sound; Echo; Air-pump; Thunder and lightning; Rainbow.
Lecture 4. Minerals, Vegetables, and Animals.
Lecture 5. Of the five senses of man – Seeing: telescope, separating colours and prisms; Hearing. Heat and Cold.

The Revised edition (1794) of this textbook, price 1s 6d, makes reference to 'air-balloons, the electrical machine, the microscope'.

In a few places there were schools that responded to the changing times and gave an education with a more practical and scientific bias. David Williams (1738–1816) not only advocated the teaching of such topics of physics as 'the general experiments of electricity, of the air pump, of statics ... the construction and use of barometers ... the principles on which heat or cold were measured ... the science of hydrostatics',[4] but practised what he preached in the school he set up in Chelsea in 1773.

A pupil at the Royal Mathematical School at Christ's Hospital, James Jurin FRS (1684–1750), introduced new subjects and himself lectured on experimental philosophy when he became master of the Newcastle Grammar School in 1709. In 1793 the Royal Grammar School of Newcastle was giving lectures three times a week on mechanics, hydrostatics, optics, sound and astronomy. Manchester Grammar School was another school that adapted its curriculum to the needs of the majority of its pupils – sons of merchants and artisans. In 1734 experiments in optics with the use of a telescope were frequent at the school.

However, Priestley's own school was not so accommodating to the needs of the times. During his childhood the young Priestley was often ill, forcing him to become an avid learner on his own. He was later sent to a private school run by a dissenting minister. Priestley writes:

> I attended two days in the week upon Mr Haggerstone, a dissenting minister in the neighbourhood, who had been educated under Mr [Colin] McLaurin. Of him I learned ... various branches of Mathematics, theoretical and practical. And at the same time I read, but with little assistance from him, Gravesande's *Elements of Natural Philosophy.* [5]

The Dissenting Academies

When the Act of Uniformity was renewed in 1662 over 2000 non-conformist rectors and vicars resigned. In order to eke out a living and provide their children with university education many of these men of learning set up schools and academies throughout the country. At Oxford students were required to subscribe to the Thirty-Nine Articles and to take the Oath of Supremacy, whilst at Cambridge they had to declare themselves members of the Church of England before proceeding to certain degrees. The early academies endured a rather makeshift existence and whether science was taught depended on the taste of the sole tutor. After the 1689 Act of Toleration better equipped academies came into existence and these usually taught science and possessed scientific apparatus.

During the century physics was taught in the Academies at Taunton, Shrewsbury, Bridgewater, Warrington, Manchester, Moorfields, Hoxton, Findern, Nottingham, Kibworth, Morpeth, Northampton, Kendal, Warwick, Carmarthen, Hoxton Square (London), Daventry, Exeter, Bristol, Abergavenny, Hackney and Rotherham. Irene Parker and H. McLachlan provide evidence that at least 26 out of 79 dissenting tutors were teaching science at the end of the century.[6] Of course, since many of these divines had been subjected to the 'grand old fortifying curriculum' of the classics, the natural philosophy taught in the early academies tended to be more Aristotelian than experimental. But some were more progressive. Thus at Sheriffhales Academy (1633–97) 'the students were conducted through a course of lectures ... in physics ... In natural philosophy, the authors read and explained were Heerebord, Magirus, De Cartes' *Principia*, Rhegius, Rohault, and De Stair, for both old and new physics.'[7]

The Cartesian Thomas Rowe (1657–1705), head of the first Newington Green Academy (1666–1706) 'was the first to desert the traditional textbooks, introducing his pupils about 1680 to what was known as "Free Philosophy".'[8] The second Newington Green Academy was conducted by Charles Morton (1627–98), whom Samuel Wesley described as an 'ingenious learned man whose chiefest excellence lay in Mathematicks and especially the mechanick part of them.'[9] Wesley goes on to state that the Academy possessed a 'laboratory and some not inconsiderable rarities with air pump, thermometer, and all sorts of mathematical instruments'. Morton wrote a textbook *Compendium Physicae* which for long remained a standard work. Owing to persistent attempts to prosecute him under the Clarendon Code and also for breaking the Stamforth Oath by giving university teaching outside Oxford and Cambridge, Morton emigrated to America, where he became Vice-Principal of Harvard College.

Natural philosophy, embracing the physics of Sucier, Rohault and Clarke, was also taught by Richard Frankland (1630–98) at his Rathmell Academy (1669–98). A pupil of Frankland, John Chorlton (1666–1706) opened the first Manchester Academy in 1699 and included natural philosophy in his course.

The physics textbooks of Le Clerc and of Du Hamel were in use in Shrewsbury Academy (1680?–1715?) and in Oswestry Academy. The physics of Le Clerc was also taught at Bethnal Green Academy (1680?–1708?) by John Ker (1623–1723) who had learnt his physics at Leyden University.

Samuel Benion (1673–1708), principal of the Broad Oak Academy in Flint, 'was a great Master in Natural Philosophy ... he found means to acquaint himself with the modern discoveries and improvements in that'.[10]

By the beginning of the 18th century the changing social and economic conditions resulting from the growth of science, industry and commerce, encouraged many dissenting academies to promote mathematics and the physical sciences with almost feverish zeal. In some 'science' came to be the most important subject taught, the academies having on their staff tutors who were eminent in their own right as scientists, men such as John Eames, Abraham Rees, John Dalton and Priestley himself.

John Eames FRS, (d.1744), the friend of Newton and who had 'a profound knowledge of mathematics and natural philosophy'[11] taught mathematics and physics at the Moorfields Academy. Eames was succeeded by David Jennings (1691–1762) who continued his distinguished predecessor's scientific work by 'encouraging the study of physical science'.

Abraham Rees FRS, (1743–1825), took his students at Hoxton Academy (1701–85) through a comprehensive course embracing mechanics, statics, hydrostatics, optics, applied mathematics and astronomy. Rees was later tutor at Hackney College. The four-year course at Findern Academy (1710–54) included natural philosophy from Le Clerc's *Physics*, from Rohault's *Physics* and from Gravesande's *Natural Philosophy*.

During the third and fourth years at Kibworth Academy (1715–23) the students went through a course of mechanics, hydrostatics, and the *Physics* of Le Clerc and of Rohault. The instruction was given by John Jennings who in 1721 published a textbook, *Miscellanea in usum Juventutis Academica*, which covered an immense range of subjects, including mechanics, hydrostatics, pneumatics, fortification and astronomy. This was the textbook used for mechanics. The course on hydrostatics comprised an abridgement of John Eames's lectures. One of the students here was Philip Doddridge who later became principal at Northampton Academy.

At the Academy between 1729 and 1751 'every student had to pay, on entrance ... a third (guinea) for the wear and tear of the scientific apparatus.' The physics instruction consisted of 'a system of natural and experimental Philosophy, comprehending Mechanics, Statics, Hydrostatics, Optics, Pneumatics, and Astronomy'. Moreover, 'this system was illustrated by a neat and pretty large philosophical Apparatus.'[12] It was here that Philip Doddridge broke away from the university and academic tradition by substituting English for Latin as the language of the lecture room.

When Doddridge died of consumption in 1751 the academy closed and the pupils were sent to a new academy established at Daventry under a former pupil of Doddridge, Caleb Ashworth.

By 1752 Priestley's health had improved. As he was debarred from the universities of Oxford and Cambridge he was sent, aged 19, to this new academy at Daventry. Because of his high standard of education, Priestley was excused the first year of the four-year course at Daventry.

The curriculum at Daventry was modelled on that at Northampton. The principal of Daventry Academy was Dr Caleb Ashworth (1722–75). Still extant (at Manchester College Library, Oxford) are 'Lecture Notes on Electricity' taken down from Dr Caleb Ashworth's lectures at Daventry Academy (1752–89). In 1770 the

Coward Trustees voted five guineas to Ashworth for improving the electrical apparatus at the academy. In 1781 the apparatus at the academy was 'very complete in Mechanics, Hydrostatics, Pneumatics, Electricity and the Airs', though 'it was defective in the Optical and Astronomical Departments.'[13] Of the 274 students who passed through the academy at least 17, including Priestley, became tutors at academies, whilst many others became schoolmasters, and so with their knowledge of physics helped in the further diffusion of knowledge of the subject.

On leaving Daventry, Priestley became a dissenting minister, first in Needham Market in Suffolk, then later at Nantwich, Cheshire. There was a need to augment his income and, like so many of his ministerial colleagues, Priestley decided to open a school. Especially after the foundation of the Royal Society, dabbling in science became almost the fashion among the higher orders of society, and, as a result, many parents wanted for their sons not only a classical education but also one that included the new experimental sciences. Many of the scientists of the time augmented their incomes by giving such tuition to the sons of the wealthy. In course of time even the middle classes demanded for their sons such an education as was more fitted to the requirements of a society becoming increasingly industrialized. That there was no lack of well-qualified tutors is evidenced from Vicesimus Knox's remark that 'happily the nation abounds with men deeply versed in Natural and Experimental Philosophy'.[14] Priestley offered to teach the classics, mathematics, etc., for half a guinea a quarter and to board pupils for 12 guineas a year. Not a single pupil came. In despair, Priestley advertised 12 lectures for adults on the use of globes. He attracted ten students who paid half a guinea a head 'which did something more than pay for my globes'.

Priestley was sufficiently encouraged to plan further lectures, but his move to Nantwich in 1758 intervened. There, despite that 'grave aversion to the business of a schoolmaster' which he shared with 'most other young men of a liberal education', he opened a school and found that 'in this employment, contrary to my expectations, I have found the greatest satisfaction'.

The school attracted over 30 pupils, six of them girls who were taught in a 'separate room'. What was notable was Priestley's teaching of science. He bought various pieces of apparatus such as an air pump and a static generator for electrical demonstrations. In his *Memoirs* Priestley tells us (1806 edition) that:

> These I taught my scholars in the highest class to keep in order, and make use of, and by entertaining their parents and friends with experiments, in which the scholars were generally the operators, and sometimes the lecturers too, I considerably extended the reputation of my school.[15]

After three years at Nantwich Priestley's success as a teacher led to his appointment in 1761 as tutor in language and literature at Warrington Academy in Lancashire, the most famous of all the dissenting academies, one which won for Warrington the title of the 'Athens of the North'.

As is evidenced from the list of eminent tutors of natural philosophy at Warrington Academy (1757–86), the physics taught at this academy was of a high order. The *Monthly Repository*, Vol. VIII, 1814 says of John Holt, the tutor 1757–72, 'in his experiments in natural philosophy he is said to have been accurate and successful, clear in his illustration of the principles to be deduced from them.'[16]

In 1760 'capital instruments were purchased at a cost of £100' in order to augment the collection of science apparatus. The Library at the academy contained no less than 81 titles in the mathematical and physical sciences.

The academy offered a three-year course for those who were intended for 'a life of Business and Commerce' as well as a five-year course for intending ministers. As at Northampton and Daventry, mathematics and natural philosophy were well represented in the curriculum. During the years 1761–7 Priestley gave lectures in physics and chemistry at Warrington, as part of the new courses he developed for students preparing for careers in industry and commerce. By emphasizing history, science and arts rather than the classics of the university curriculum he sought to prepare students for practical life. Where textbooks were not available Priestley wrote these himself. In his *Essay on a Course of Liberal Education for Civil and Active Life* (1765) he dismissed the contemporary grammar school curriculum as 'a common topic of ridicule' and argued for a fundamentally new curriculum which would seek to satisfy three main needs: religious, intellectual and utilitarian. He proposed as essential subjects Latin, English, French, mathematics, physics, chemistry, history and geography. Howson points out that this curriculum was adopted over a century later in grammar schools and was, in fact, exactly the same as the curriculum he followed when taking his School Certificate examinations (equivalent to GCSE) in the 1940s.[17] Priestley continued by arguing that unless such changes were made 'we must not be surprised to find our schools, academies and universities deserted, as wholly unfit to qualify men to appear with advantage in the present age.'

It was while he was at the Academy that Priestley wrote his *History and Present State of Electricity* and carried out the experiments which gained him admission to the Royal Society in 1766. In 1768 he published *A Familiar Introduction to the Study of Electricity*, a textbook for beginners, in the preface of which he pointed out the necessity for practical experimental work in studying a subject such as electricity.

Priestley also furthered the cause of experimental physics teaching by his persistent advocacy of the subject in his treatises on education. In the Preface to his *Experiments and Observations relating to Various Branches of Natural Philosophy* (1779) he urges that:

> if we wish to lay a good foundation for a philosophical taste, and philosophical pursuits, persons should be accustomed to the sight of experiments, and processes, in early life. They should, more especially, be early initiated in the theory and practice of investigations, by which many of the old discoveries may be made to be really their own; on account of which they will be much more valued by them.[18]

After six years at Warrington, Priestley became minister at Mill Hill Chapel, Leeds. His interests from that time onwards turned more and more to chemistry. Yet even so he was still to make significant contributions to adult education – he helped, for example, to establish a circulating library in Leeds – and to the dissemination and popularization of scientific knowledge. When he later moved to Birmingham he was an active member of the Lunar Society, so-called because the Society met every full moon. It was also called the Lunatic Society. It was anything but the latter, counting among its members Boulton and Walton, who were close friends of Priestley. The Society, which flourished between 1766 and 1791, actively promoted science and its application to industry and crafts.

On 14 July, 1791, the second anniversary of the storming of the Bastille prison in Paris, riots broke out in Birmingham. Because of his open sympathy with the French (and American) Revolutions, the mob burnt down Priestley's church, house, library and laboratory. He was lucky to escape with his life. For the next two years Priestley taught at New College (Hackney College) in London. He gave lectures on experimental philosophy, which included the subjects of heat, light, electricity and magnetism. In 1794 he emigrated to the United States.

Priestley was not alone in the dissenting academies in teaching physics experimentally. Other physics teachers such as Dr William Enfield, Caleb Rotherham, John Horsley, John Collet Ryland and John Dalton thought such methods essential.

Dr William Enfield, like Priestley, taught at Warrington Academy. He published two physics textbooks, being the author of *Institutes of Natural Philosophy, Theoretical and Experimental Physics* (1785) and the editor of *Scientific Amusements in Philosophy and Mathematics* (1821).

At Kendal Academy Caleb Rotherham (1694–1758) 'not only instructed his pupils in the theory of mathematics and natural philosophy, but also possessed the happy talent of illustrating them with great success, by means of experiments performed with an extensive, and, for that time, well constructed apparatus'.[19]

Much of his apparatus Rotherham had acquired from Morpeth Academy, where John Horsley had taught 'a complete course of experimental philosophy', covering statics, mechanics, hydrostatics and pneumatics, which were illustrated by experimental demonstrations.[20]

In 1792, John Collet Ryland (1723–92) who included natural philosophy as a subject for study in his academy first at Warwick and then at Northampton, published his *Address to the Ingenious Youth of Great Britain*. In this, Ryland pointed out that lack of apparatus was no justification for teaching physics merely out of books. As he said:

> ... a fire shovel, tongs and poker, will shew the foundation of the mechanic powers; especially the nature of levers. A spinning wheel will clearly shew the power of the wheel and axle ... marbles, will teach a schoolboy the nature of percussion and the laws of motion ... The twirling of a chamber maid's mop will show the nature of the centrifugal force of the planets. The fall of a farthing ball teaches the doctrine of gravitation and the laws of falling bodies ... A sponge will teach the rise of water in capillary tubes. A syringe, or a squib, or sucking with a reed, or a wheaten straw, will shew the nature of pumpwork. A schoolboy's jew's harp, will serve to teach us those tremulous motions which are the cause of sounds; and a glass prism and soap bubbles, a looking glass and an ox's eye from the butcher's, will be a happy foundation for optics.'[21]

At Manchester Academy (1786–1803) the tutor in mathematics and natural philosophy was none other than the celebrated Quaker scientist John Dalton (1760–1844). When the Academy moved to York in 1799 Dalton taught mathematics and natural philosophy privately (in Manchester) at a charge of 18 pence an hour.

That the physics instruction given at this academy included experimental work is evidenced by the numerous references to the purchase of apparatus. In March 1792 the scientific apparatus and other furniture was insured for £200.[22]

Private Schools and Academies

It was not only the Dissenting Academies that gave instruction in physics. Hans has adduced evidence to show that private academies with a scientific, technological curriculum abounded in England during the 17th and 18th centuries.[23] These catered mostly for the lower-middle classes. Some of these academies were large establishments with well-equipped laboratories.

Academies in London

Hackney Academy was the oldest and largest of these, and could boast two headmasters who were Fellows of the Royal Society. Henry Cavendish was a pupil here. Physics was also taught by John Harris (1666–1719), the author of the *Lexicon Technicum*, at his academy in Amen Corner.

At Little Tower Street Academy (1715–60), Benjamin Worster, assisted by Thomas Watts, lectured on physics. From Worster's *Principles of Natural Philosophy*, published in 1722, we learn that his lectures were illustrated by means of experiments 'in the mechanism of machines, in hydrostatics, the laws of motion, pneumatics and optics'. In 1716 Watts had translated Rohault's *Treatise on Mechanics* into English, 'at the instance and desire of that excellent mathematician Mr. Humphrey Ditton', who wanted such a textbook for his Mathematical School at Christ's Hospital.

Martin Clare included experimental natural philosophy at his Soho Academy (1717–1801), which possessed 'a large Apparatus of Machines and Instruments'. At his 'seminary' in Islington, John Evans gave instruction in physics 'by means of a well constructed apparatus'. William Nicholson (1753–1815), the author of *Introduction to Natural Philosophy*, kept a mathematical school in London from 1778.

The Royal Military Academy, Woolwich, was opened towards the end of the century. Here Dr Charles Hutton gave instruction in mechanics, hydrostatics, hydraulics, pneumatics, and sound, in the course of his duties as Professor of Mathematics. Long before his appointment to the Academy, Hutton, in 1760, opened a school in Newcastle in which he gave instruction in mechanics, statics and hydrostatics.

Among other institutions that taught physics were the Mathematical School at Wapping, the Mathematical Academy in Chelsea, the Military Academy in Little Chelsea, the City Commercial School in Lombard Street, and the academies in Bridgewater Square and in Wandsworth.

Academies outside London

Physics was also taught at York Academy, Heath (Wakefield) Academy, Cirencester, Compton (Dorset), and Stroud Academy in Gloucester.

The Rev. John Wesley, who in 1746 founded Kingswood School, near Bristol, and at which experimental philosophy was taught, wrote *Desideratum; or Electricity made Plain* in 1759. Four years later was published his popular *Survey of the Wisdom of God in Creation or a Compendium of Natural Philosophy.* This book deals with such topics as the lodestone, properties of air, rainbow, vacuum, laws of motion, light and colours, sound, heat, gravity and the heavenly bodies.

At his Academy in Bristol Benjamin Donne taught 'Hydraulics, Hydrostatics ... Mechanics ... Optics ... Pneumatics... with the use of Mathematical and Philosophical Instruments.'[24] In 1769 Donne published *An Epitome of Natural and Experimental Philosophy ... Designed for the Use of Schools.* This textbook covered mechanics, hydrostatics, pneumatics, optics, electricity, magnetism and astronomy.

At his academy in Salford, Henry Clarke (1743–1818) gave 'select lessons on the most popular parts of Natural and Experimental Philosophy, illustrated by numbers of experiments, with an elegant and extensive Apparatus, newly fitted for the purpose'.[25]

When the Manchester College of Arts and Science was founded in 1783 Clarke was appointed Prelector in Mathematics and Experimental Philosophy. Clarke's lectures, which were profusely illustrated by experimental demonstrations, covered the fields of electricity, magnetism, pneumatics, pneumatic chemistry, optics, astronomy, mathematics and mechanics.[26] In 1790, with his brother, he founded the Literary, Commercial and Mathematical School, Mount Pleasant, Liverpool.

At J.B. Florian's Academy in Bath, mechanics and astronomy were taught in the fourth year (children aged 10–11 years), and physics and chemistry in the fifth year (children aged 11–12 years). This teaching was accompanied by experimental demonstrations: more than that, the children were to 'repeat the various operations and experiments.'[27]

This brief summary shows quite conclusively that during the 18th century England abounded with private academies and schools that gave instruction in physics, suitably illustrated by means of experiments.

Schools and Academies in Scotland

Though the subjects were usually restricted to classics, some science also began to be taught in the burgh or grammar schools of Scotland during the 18th century. Several such schools, particularly in and near seaport towns, began to give instruction in navigation which included much incidental physics, such as the compass and the telescope.

The Industrial Revolution in Scotland was now beginning to make its influence felt in the sphere of education. Numerous voices were raised for the inclusion of the mathematical and physical sciences in the curriculum. In 1746, for example, John Mair represented to Ayr Town Council the need for subjects like 'Navigation... Mathematical Sciences and some parts of the Natural Philosophy', and when Mair was elected Rector of the Ayr Grammar School he taught these subjects. The fee for a course of 'Natural Philosophy comprehending the Doctrine of Attraction, Gravitation, Congress of Bodies, Mechanick Powers, Pendulums, etc., Hydrostaticks, Pneumaticks, Opticks and Astronomy, with the several Experiments' was two guineas. In 1773 payment was made of 18s 10¾d to cover the cost of 'a crystal receiver to the airpump and tubes for mathematical instruments'; and in 1776 Ayr Council paid £1 15s for 'two new wheels for the electorising machine, two vertical and horizontal axes, a pedestal for vertical axis screws and for repairing the same, furnishing a new intermitting fountain, tantalus cup to fit the side of the large conductor, and for soldering the spiers of the air-pump.'[28] George Chapman (1723–1806), the Rector of Dumfries Grammar School, also advocated physics teaching in the grammar schools.

Just as in England the traditional conservatism of the old foundation public and grammar schools had resulted in the establishment of several private academies in which a wider and more progressive education was given, so, too, in Scotland, similar academies came to be established. Perth Academy, which was founded in 1760, taught 'a course of Natural Philosophy illustrated by experiments', comprising statics, dynamics, hydrostatics, pneumatics, optics, astronomy, and chemistry (including heat). In 1779 the Rector of the academy, Alexander Gibson, bought scientific apparatus including an electric machine, a set of models of the mechanical powers, a universal microscope, a solar microscope, concave and convex mirrors, and new globes. The course of instruction at Dundee Academy was substantially the same as at Perth. Physics was also taught at Inverness Academy, founded in 1787.[29] Instruction in 'natural philosophy and astronomy, for which proper apparatus was procured' was also given at Salton Academy.[30]

The prospectus of Ayr Academy, issued in 1794, stated that masters of distinguished abilities were to be engaged for teaching mechanics, navigation, astronomy, and natural and experimental philosophy. In 1797 it was reported that 'the value of the Mathematical and Philosophical Apparatus of the best and most useful kinds, already procured amounts to at least 600 pounds.'

Physics in the Education of Girls

For those girls whose families could afford it, education was given at home by private tutors, some of whom, in the course of their tuition, might have touched on the subject of natural experimental philosophy, which was then the fashionable study. We saw that at his school in Nantwich Priestley took girls, who were taught in a 'separate room'. There were also private boarding schools, but most of these were too preoccupied with education in the gentler accomplishments to bother about teaching such 'masculine' subjects as the mathematical and physical sciences. For girls of the lower classes there was no education.

That such a narrow education was criticized is not surprising. In 1792 Mary Wollstonecraft (1759–97) issued her *Vindication of the Rights of Woman* in which she not only objected to the prevalent aim of educating women as the 'playthings' of men but also advocated the teaching of simple physics to young children. Lady Sarah Pennington also recommended study of physics in her *An Unfortunate Mother's Advice to her absent Daughters* (1761) and again in her *Instructions for a Young Lady* (1773). A similar recommendation was made by Mrs Chapone in her *Letters on the Improvement of the Mind Addressed to a young Lady* (1774). An eloquent plea for educating girls and women and for instructing them in the mathematical and physical sciences was made by Dr Erasmus Darwin (1731–1802) in his *Plan for the Conduct of Female Education,in Boarding Schools* (1797). As a result of such pleas certain girls' boarding schools did give instruction in physics. During the 1790s Mrs Margaret Bryan established such a school.[31] In 1797 she published the *Lectures on Astronomy and Mathematics* which she delivered in her boarding school. Her lectures comprised:

Lecture I	On Optics. On the progressive Motion of Light, and its Velocity. The Nature of Light. Of Refraction and Reflection – the Laws of. Of the Nature of Lenses. Of Gravity. Of Pendulums. Of the Increase of Matter about the Equator.
Lecture II	Astronomy.
Lecture III	Of the Atmosphere. Of the Magnet.
Lecture IV	The Copernican System explained.
Lecture V	Mentions the Abberation of Light.
Lecture VI	Mentions Motion and Gravity.
Lecture VII	Gravity, Weight, Attraction, Centrifugal and Centripetal Forces in relation to Astronomy. Description of the Eye.
Lecture VIII	Of the Laws of Shadow. Of Eclipses, etc.

The pupils at the school run by Miss Runciman and Miss Roberts at No. 32, Brunswick Terrace, Brighton, obtained 'the fullest instruction in science by attending a course of nine lectures by Walker in a public room in Brighton, one on Electricity, one on Galvanism, one on Optics, one on Hydrostatics, Mechanics, Pneumatics and three on Astronomy'.

In addition to the instruction in physics given in some girls' boarding schools, many women obtained considerable instruction in the subject through attending the experimental lectures and demonstrations of that numerous band of itinerant physics teachers discussed earlier.

In addition to Mrs Bryan, other women took to writing physics textbooks. Mrs Priscilla Wakefield (1751–1832) in 1794 published her *Mental Improvement*, in which she discussed 'Philosophical Instruments, Galileo, the telescope, the microscope, and astronomy'. She followed this with *Domestic Recreation; or Dialogues illustrative of Natural and Scientific Subjectes between a well-informed mother and her daughter* ... In her discussion of light and colours she refers to Newton's experiments.

Chapter Four

Physics for the Masses – the Mechanics' Institutions George Birkbeck (1776–1841)

Whilst the Scientific Institutions and Philosophical Societies catered for the more sophisticated sections of the community, there was little opportunity for the lower classes to gain an acquaintance with science. Whilst Lord Brougham and some others were alive to the need of giving the lower classes more in the way of education and enlightenment, there were others, such as Henry Petty, who argued that if the labouring classes became educated they would overthrow the Government and overturn society. That the views of the former group prevailed in providing adult education for the masses is due in no small measure to the pioneering work of the physician George Birkbeck.

Birkbeck was born in Settle, in Yorkshire, in 1776. He studied at Edinburgh University, and in 1799, at the age of 23, was appointed Professor of Natural Philosophy in the Andersonian Institution in Glasgow.

The Andersonian Institution

Regular courses in more modern physics had begun at Glasgow University with the appointment in 1727 of Robert Dick as Professor of Natural Philosophy. In that same year the University enacted that any private citizen, though not a student, might attend classes in physics 'without a gown'.

In 1751 Dick's son succeeded him as Professor. In addition to his university classes the younger Dick also held an evening class in Experimental philosophy comprising mechanics, hydrostatics, pneumatics, heat, optics and astronomy. When the younger Dick died in 1757 he was succeeded by John Anderson who founded what came to be known as the Andersonian Institution. Like his predecessors, Anderson continued the evening classes in Physics. But during his time the class of mechanics, artisans, tradesmen and townsfolk reached almost unmanageable proportions. His course on two evenings a week was of a more popular nature and without mathematics but illustrated by experiments. In this 'anti-toga' class as he called it, the students would get a lecture and then move to an 'experiment room' to see experiments performed.

When Anderson died in 1796 he bequeathed his property to establish the Andersonian Institution, which was to supply courses of general and scientific instruction open to all classes and to both sexes. Anderson did much to encourage science teaching for women. In his will he stipulated that 'at least once a year there would also be a Ladies' Course of Physics Lectures'.[1] This came to be astonishingly successful. In its first session in 1796–7 a total of 792 people attended, half of whom were women. Unfortunately Anderson's property only amounted to £1000 and so his trustees had to be content with establishing only a Chair of Physics.

Thomas Garnett (1766–1802) was appointed to this Chair of Natural Philosophy. Like Anderson, Garnett had great success with his adult education courses in experimental philosophy. In 1799 Garnett was appointed to the Royal Institution in London, and was succeeded by Birkbeck.

Birkbeck continued the tradition of giving lectures in physics to mechanics and artisans. He needed apparatus and had to instruct the workmen how to make it. He was struck by the zeal with which they listened to his directions and with the 'intelligent curiosity' they showed in the working of a modern centrifugal pump. He invited some of them to attend his lectures. Lack of space led him to propose lecturing to the artisans separately. But his committee objected, 'If invited, the mechanics would not come; if they came, they would not listen; if they listened, they would not comprehend.' Birkbeck insisted; they came, they listened, and they comprehended.

His first course began with 70 students. The numbers were soon to rise to well over 400. In fact the numbers got so large that Birkbeck had to give his lectures in separate classes in a building known as Anderson's Institute. In 1800 he gave free lectures on physics to 500 mechanics. These lectures abounded with experiments and were conducted 'with the greatest simplicity of expression and familiarity of illustration'.

It was courses such as these that were to lead to the foundation of the Mechanics' Institutions.

Birkbeck resigned in 1804 and settled in London as a physician. He was succeeded in the Andersonian Institution by Andrew Ure, later FRS, who occupied the Chair from 1804 till 1830. Ure continued the tradition of evening classes addressed to 'the Manufacturers and operative Artisans'. This course, consisting of the elements of Mechanical Science, 'explained in the simplest and most impressive manner', was given to a class consisting of about 500 men. These lectures continued right up to the time of the formation of the Glasgow Mechanics' Institution in 1823.

Among the prominent students of the Andersonian Institution were William Thomson (later Lord Kelvin) and his brother James, who became Professor of Engineering in the University of Glasgow, and Lyon Playfair who became Professor of Chemistry at Edinburgh University in 1858.

Mechanics' Institutions

The Industrial Revolution had produced a large mass of artisans and factory operatives who thirsted for scientific knowledge. It was in an endeavour to satisfy this thirst that the system of evening classes and Mechanics' Institutions grew up. As early as 1772, for example, Henry Clarke had started a night school for adult mechanics at Salford. The Edinburgh School of Arts was opened in 1821, offering a diploma to those students who had attended a three-year course and successfully passed an examination in mathematics, physics and chemistry. Students were encouraged to carry out experiments themselves. As D.B. Reid, a lecturer in physics and chemistry reported, 'several of them performed nearly two-hundred'.[2] Both here and at the Watt Institution in Edinburgh all the teaching was done by evening lectures. In 1868, giving evidence to the Select Committee on Scientific instruction, Dr S. Macadam stated that for the past 12 years he had been giving popular lectures on science for the working classes in the evenings at the School of Arts and that the attendance occasionally rose as high as 1400.

In 1823 there was a dispute at the Andersonian Institution in Glasgow between the management and the mechanics about the use of the library, and Birkbeck's advice was sought. As a result the mechanics seceded and formed their own Glasgow Mechanics' Institution, the first of its kind, with Birkbeck as honorary patron. In two years the Institution had 1200 students and a library of 1639 volumes. The students appointed the lecturers and re-elected them annually. In course of time a more normal type of organization was adopted and appointments were made by a committee.

In the following year Birkbeck proposed the foundation of a London Mechanics' Institution of which he remained president until his death. In 1907 this institution was renamed Birkbeck College and in 1920 it was recognized as a school of the University of London for evening and part-time students.

The success of the London Institution led to the establishment of similar institutions all over Britain, some of which later developed into technical colleges. In 1824 an Institution was founded in Manchester 'to give instruction in chemistry and physics and to expound scientific principles applied in factories'. Between 1835 and 1839, 235 lectures on physical science were delivered at the Manchester Mechanics' Institution. The average attendance at each of the 65 lectures given in 1838–9 was 4000. In 1836–7 student-teachers were admitted to the evening classes of the Liverpool Institution. In 1839 there were 3478 members of the Institution, 473 of whom were ladies.

By 1850 there were 610 Mechanics' Institutions with 120 000 registered members. Detrosier remarked in 1829 that:

> In these excellent institutions ... the mechanic is taught ... mathematics ... the laws of the simple mechanical powers are explained to him, together with the nature of complex motions, the doctrine of central forces, and the application of the whole to purposes of commerce ... From the lectures on Pneumatics he learns the nature of the mechanical properties of air, the value of atmospheric pressure as a moving power, and he is taught the principle on which the first steam engines were constructed ... by the principles of Hydraulics the mechanic is taught to apply water to other and most important uses ... It is the peculiar feature of modern times that the doors of the temple of science have been thrown [open] to all comers.[3]

The Annual Report of the Leeds Mechanics' Institution, 1856, shows that 21 lectures on science had been given. The Report goes on to say that:

> it has long been their [the Committee's] wish that the principles of Natural Science should be taught to the working man – not by Catechisms or dogmatic instruction, but by placing before him a series of well-arranged experiments,

which should demonstrate to him the laws which govern the natural world; while at the same time he is going through a course of Mathematical instruction which enables him fully to comprehend the problems presented to him.[4]

In addition classes on experimental natural philosophy were given to men in the evenings.

In 1857, while Professor of Natural Philosophy at Aberdeen University, the great physicist James Clerk Maxwell ran a class on Monday evenings for working men. Clerk Maxwell commented on the 'operatives who do better exercises than the University men about false balances, quantity of work, etc.'

Many of the lectures on science were given by eminent teachers and scientists, including University Professors. That hard-working artisans, after a full day's labour, were prepared in large numbers to attend lectures on science is proof of the high regard in which science was held. As the *Edinburgh Review* commented: 'the sacred thirst for science is becoming epidemic.'

Often Mechanics' Institutions would combine together for the purpose of interchanging lecturers and books and exhibits. In order to assist itinerant lecturers the Working Men's Educational Union prepared sets of diagrams ready for use: for example, diagrams of a telescope etc. Groups of mechanics with similar interests in a particular science often formed mutual instruction classes. This mutual instruction often included the repetition of experiments by the mechanics themselves.

Most Institutions possessed a considerable stock of physics apparatus, and many, such as the Institutions in London, Leeds, Bradford, York, Manchester, Liverpool, Coventry, Newcastle, Huddersfield and Keighley, possessed laboratories which were used with great effectiveness.[5]

The following is a brief selection from the list of apparatus possessed by the Liverpool Mechanics' Institution:[6]

Mechanics	Atwood's machine, inclined plane, wheel and axle, etc.
Hydraulics	Force and lift pump, Archimedean screw, etc.
Hydrostatics	Rotary blowing engine.
Pneumatics	Air pumps, guinea and feather apparatus, brass hemispheres.
Optics	Magic Lantern.
Electricity	Cylindrical machine, battery of five jars, etc.
Galvanism	Troughs and piles.
Chemistry	Cavendish's eudiometer, Davy safety lamp, Leslie's freezing apparatus.

Many of the more progressive Mechanics' Institutions, such as those at Manchester, Liverpool, Birmingham and Leeds, established schools attached to them in which elementary science was also taught. The Birmingham and Midland Institution, founded in 1853, maintained a School of Industrial Science, in which chemistry, mechanics, and geology were each taught with a practical bias. The Leeds Mechanics' Institution established a day school in 1854, in which experimental natural philosophy was taught. A girls' school was established the following year, also with instruction in experimental philosophy forming part of the curriculum.

Mechanics' Institution schools were opened in Salford in 1853 and in Manchester in 1857. These schools were both developed by Mr Angell, who later became science master at Manchester Grammar School. The object of the schools was 'not so much to teach as *to inspire or develop a taste for science*' [my italics]. Here we have a shift of emphasis away from the utilitarian view of science teaching.

The Society of Arts

In 1851 Hugo Reid recommended the conversion of Mechanics' Institutions into colleges for the people. He proposed a four-year course of systematic instruction in mathematics, physical and biological science, and mental and moral philosophy. At the same time, on the initiative of the Rev. James Booth, a small committee of the Society of Arts was set up. The Committee stressed the importance of physics and chemistry in the improvement of manufactures and went on to urge 'in view of our educational backwardness and the enterprise shown by our foreign rivals in technical education', that Mechanics' Institutions should be converted into industrial schools, with set syllabuses, examinations and prizes. It recommended the establishment of a central technical college either in London or in Manchester, to teach and to examine in science. In 1853 the Society of Arts formed a union of 400 leading Mechanics' Institutions with a view to improving standards. It proposed to do this by a comprehensive system of science examinations. But this step evidently frightened off the artisans. In 1855 only one candidate took the examination, which was intended for persons over 15 who had already left school. The syllabus was suitably modified and in 1856 there were 52 candidates. In 1862 Sophia Jex-Blake became the first successful woman candidate. In 1856 Booth wrote 'How to Learn and What to Learn' in which he showed mechanics and artisans why, what and how they should learn. Booth's ideas were disparagingly spoken of by the Athenaeum as the 'Knowledge-box plan', but the action of the Society of Arts in promoting examinations awakened the universities and led in course of time to the setting up of the Local Examinations of Oxford and Cambridge.

As the century wore on there was a tendency for the lectures on the physical sciences to become less popular. This was due to many mechanics losing heart when they found that their poor basic education set a limit to what scientific knowledge they could acquire. As other centres of scientific education became available, the Institutions tended to decline as places where artisans and operatives could acquire some knowledge of science. Instead they became little more than 'an association of shopkeepers and their apprentices, law copyists, and attorneys' clerks', whose interests lay more in literature and in the social amenities offered by the Institutions.[7]

Working Men's Colleges and Co-operatives

Elementary science formed part of the curriculum of the Working Men's Colleges which came to be built to make good the deficiencies in the education of the working classes. These Colleges had their origin in the Chartist and Christian

Socialist movements. The first People's College was established at Sheffield in 1842. The College was open to those over 16, and had a wide curriculum, the classes being held in the early hours of the morning or in the evening. A Working Men's College was founded in London in 1853. Science had an important place in the curriculum. At Halifax Working Men's College, founded in 1855, instruction was given in evening classes in chemistry, magnetism, electricity, light, applied mechanics and machine drawing. In 1868 evening lectures in physics were given to working men at the Royal School of Mines in Jermyn Street, London. In his evidence to the Select Committee on Scientific instruction in 1868 Professor Tyndall stated:

> The working men come there night after night. I have sometimes to lecture before another audience, at the Royal Institution, composed of what you would call a higher stratum of society; but, with regard to gentlemanly demeanour and courteous attention, I cannot conscientiously say that I prefer one of those audiences to the other. I have purposely looked round the filled benches in search of a yawn; I never once saw it among these working men. I often receive letters which are perfectly touching, in the name of 20 workmen, or 30 workmen, in such and such a factory, expressing their intense disappointment at not being able to get tickets for the lectures.[8]

Science was taught in many of the Co-operative Societies that flourished throughout the country in the second half of the 19th century. In Blaydon, Northumberland, the Society had 260 students studying mechanics and mathematics. At Sunderland 107 students attended classes in steam, applied mechanics and machine construction. The Skelton, Yorkshire, Society organized Science and Night Schools. Bolton had 300 students in its ten science and two technical classes. At Pilkington 69 students were studying machine construction, mathematics, magnetism, electricity, chemistry, cotton manufacture, among other subjects. At Manchester Equitable Society 129 students studied machine construction, applied mechanics and steam. The Rochdale Equitable Pioneers had 490 workmen studying machinery, mechanics, mechanical engineering, chemistry, mathematics, agriculture, biology. The Ramsbottom Industrial Society in Rossendale district had 161 workmen studying machinery, mathematics, sound, light, heat, magnetism and chemistry.

Private Lecturers and Instrument-makers

The private lecturers of science, who explained and demonstrated the principal topics of physics to appreciative audiences, also played a very important role in the general diffusion of scientific knowledge. Apart from those many lecturers who have already been mentioned as giving courses on physics either inside formal educational institutions or outside them, there were many others who thus helped in diffusing knowledge in physics. The newspapers of the time, such as the *Manchester Guardian*, for instance, abound with advertisements relating to courses of lectures on physics. *The Times* of 9 November 1815 mentions 'the peripatetic philosophers, who give lectures for a guinea each course, in every village, near London'. In 1841 Thomas Coates listed 45 such lectures in physics.[9]

In 1815 C.H. Wilkinson (1765–1844), who described himself as a 'Lecturer on Galvanism in Soho-Square and Lecturer on Experimental Philosophy to the City Institution', was giving his 'Course of Experimental Philosophy' at Bath. Richard Dalton gave courses on physics in various towns of the country. In 1816 he was giving a course of nine lectures on mechanics, hydrostatics, hydraulics, and optics at York. William Sturgeon, who made fundamental discoveries in electricity, travelled from town to town by horse, a cart conveying his science apparatus.

During 1823 and 1824 a Mr Jackson gave lectures on physics in Settle. William Paley, the schoolmaster of Giggleswick, who attended Jackson's lectures, noted in his diary that 'many of us were electrified by putting each hand into a bason [sic] of water into which he put a small wire'. Some other lecturers in physics were George Singer, of the Scientific Institution, Cavendish Square, Mr Hardie of 97 Pall Mall, and Mr Nield of Clitheroe.

The **scientific instrument-makers** should not be excluded from those who furthered the teaching of physics. Many of these instrument-makers also lectured on physics. Among the London instrument-makers were George Adams, Jesse Ramsden, William Carey, Edward Troughton and W. Simons, James Smith and R.L. Beck, Andrew Ross, Hugh Powell, W.& S. Jones, W. Newberry, Tagliabue and Cassella, Henry Negretti and W. Zambra.

At the back of Richard Potter's *An Elementary Treatise on Mechanics* (1846) Taylor and Walton, of 28 Upper Gower Street, advertised their 'Educational Models for the use of schools, Mechanics' Institutions, and for private "instruction".' The following is a selection from their list.

> Apparatus for cohesion, capillary attraction, electric, and magnetic attraction, impenetrability and inertia, with descriptions and diagrams. Price 21sh. in a box. A machine for illustrating centrifugal motion 10sh. Atwood's machine for explaining the laws of falling bodies £2.2sh.
> Sets of Mechanical Powers.
> A Double Inclined Plane, with an application of the composition and resolution of forces 10sh.
> A set of apparatus for hydrostatics, hydraulics and pneumatics £6.6sh.
> Sectional model of Steam Engine £2.2sh.
> Pyrometer, for showing expansion of metals 15sh.

In the same book appeared an advertisement for large, coloured 'Mechanical Diagrams for the use of Lecturers, Philosophical Classes, and Schools'.

Textbooks

During the first half of the 19th century a large number of physics textbooks were published, ranging from extremely simple children's books to learned treatises. Many of these textbooks have already been mentioned. The very number of these books and the fact that many of them went into several editions reflects the great demand that existed for knowledge of the physical sciences.

In 1823 appeared the first number of the *Mechanics' Magazine*, costing three pence. This magazine, which Birkbeck described as 'the most valuable gift which the hand of Science has yet offered to the artisan', treated of colour, electricity, engine work, galvanism, heating and ventilating, magnetism, mechanical powers, optics and optical instruments, perpetual motion, pumps and the steam engine. The engravings, of which there were over 700, illustrated such things as the diving bell, steam engines, Leslie's apparatus for making ice, Hero's steam engine, Watt's single reciprocating engine and the steam forge hammer.

At the instigation of Lord Brougham the Society for Diffusion of Useful Knowledge was founded in 1826 to publish cheap and simple textbooks for the lower classes. The Society was ridiculed as the 'Sixpenny Science Society' and as the 'Steam Intellect Society'. Between 1829 and 1838 the small treatises that comprised the Library of Useful Knowledge and which covered the whole field of physics, came to be published. Moreover, these cheap and well-illustrated books were written by eminent experts such as Lardner, Millington, Brougham, Sir David Brewster, P.M. Roget and others.

The great popularity of these textbooks is shown by the fact that within nine months of the appearance of the first treatise a circulation of nearly 20 000 had been secured for each number. Though it was the intention to cater for 'the million' these treatises were, in fact, much too advanced for the working classes. They were of the standard of the present General Certificate of Secondary Education and in many cases they reached GCE 'Advanced' Level, as, for instance, in discussing double refraction, polarization, galvanometers, and the like.

The Society also published a *Penny Magazine* and a *Penny Cyclopaedia*, both of which contained miscellaneous information, including science. In 1840 the Society considered the issue of apparatus to help the study of science further. Between 1842 and 1843 it issued an excellent *Series of Diagrams illustrative of the Principles of Mechanical and Natural Philosophy*. These large coloured wallcharts can be seen in the British Library.

In this chapter we have surveyed the great efforts made to bring science within the reach of the working class. Those efforts were furthered by the many cheap, simple-to-understand books and booklets which came to be published. It was an apprentice who eked out a living binding such books together who was to make the next great leap in the advance of physics and physics teaching. The name of the apprentice bookbinder was Michael Faraday.

Chapter Five

The Royal Institution
Michael Faraday (1791–1867)

Michael Faraday, one of the greatest men of science of any time, the third son of a blacksmith who had migrated from Yorkshire to London soon after marriage, was born in London on 22 September, 1791. He had no education as such. After some attendance at a school of the simplest description he was apprenticed at the age of 13 to a small bookseller and bookbinder. At first he delivered newspapers to his master's customers. At the age of 17 he was allowed to learn bookbinding. He avidly read any books that came into his hands for binding, especially those on chemistry or electricity.

Faraday reminisced later:

> Whilst an apprentice I loved to read the scientific books which were under my hands, and amongst them, delighted in Marcet's *Conversations in Chemistry*, and the electrical treatises in the *Encyclopaedia Britannica*. I made such simple experiments in chemistry as could be defrayed in their expense by a few pence per week, and also constructed an electrical machine, first with a glass phial, and afterwards with a real cylinder, as well as other electrical apparatus of a corresponding kind. ... My master allowed me to go occasionally of an evening to hear the lectures delivered by Mr. Tatum on natural philosophy at his house, 53 Dorset Street, Fleet Street. The charge was one shilling per lecture, and my brother Robert (who was three years older) made me a present of the money for several. I attended twelve or thirteen lectures between February 19, 1810 and September 26, 1811. It was at these lectures I first became acquainted with Magrath, Newton, Nicol, and others.[1]

Faraday was fired with zeal for science:

> The desire to be engaged in scientific occupation, even though of the lowest kind, induced me, whilst an apprentice, to write, in my ignorance of the world and simplicity of my mind, to Sir Joseph Banks, then President of the Royal Society. Naturally enough, 'no answer' was the reply left with the porter.

One day one of his master's customers gave the young Faraday tickets to attend Sir Humphrey Davy's lectures on chemistry at the Royal Institution. The course was the last Davy delivered as Professor, and Faraday made careful notes of everything he heard and saw. In his spare time Faraday neatly wrote out the lectures given by Davy and illustrated them with drawings in a quarto notebook. This he sent to Davy, at the same time requesting any kind of job in his laboratory. Davy sought the advice of one of the governors of the Royal Institution. 'Let him wash bottles,' said the governor. 'If he's any good he'll accept the work; if he refuses, he is not good for anything.' Davy interviewed the young Faraday and advised him against a scientific career, telling him science was a hard mistress. Faraday went back to bookbinding.

Then fate played a hand. One evening Mr Harris, the Librarian and Superintendent of the Royal institution, heard a commotion in the lecture room. He went to investigate, and found that over an argument regarding the needs of Professor Brande, the Professor of Chemistry, Mr Payne, the laboratory assistant, and Mr Newman, an instrument-maker, had come to blows. Though Mr Payne seemed to have got the better of the argument physically, he got the sack. Some months later, on 1 March 1813, Sir Humphrey Davy reported to the Managers of the Royal Institution as follows:

> Sir Humphrey Davy has the honour to inform the Managers that he has found a person who is desirous to occupy the situation in the Institution lately filled by William Payne. His name is Michael Faraday. He is a youth of twenty years of age. As far as Sir H. Davy has been able to observe or ascertain, he appears well fitted for the situation. His habits seem good, his disposition active and cheerful, and his manner intelligent.'[2]

Faraday thus began his career at the Royal institution, at a salary of 25s. per week, with two rooms at the top of the house. In due course Faraday was to succeed Davy as Professor; and to raise the Institution to the pinnacle of scientific fame, with huge crowds, including royalty, flocking to Albemarle Street to hear his lectures. Faraday spent his whole working life, from 1813 to 1858, at the Royal Institution.

The Scientific Societies

The Royal Institution was one of many scientific societies founded during the 18th and 19th centuries.

In 1766 the Lunar Society of Birmingham was founded. This Society, which included among its members many famous scientists, such as Erasmus Darwin, Priestley, James Watt, Sir Joseph Banks and William Herschel, did much to spread knowledge of science among the people of the Midlands.

The Manchester Literary and Philosophical Society was founded in 1781. One of its most distinguished members was John Dalton. The publication of the Society's *Memoirs* did much to spread scientific knowledge.

The London Institution was founded in 1805 to promote 'the diffusion of science, literature and the arts' by means of lectures and experimental demonstrations. Between 1807 and 1812 £2482 was spent on apparatus. In 1831 Birkbeck claimed that the apparatus was 'most perfect in the branches of pneumatics, hydrostatics, common and voltaic electricity, and electro-magnetism'. The physics lectures were given by very able and well-known teachers, men such as Birkbeck, Millington, the Professor of Mechanics at the Royal institution, Ritchie and Lardner, both Professors at the University of London. In the 1840s W.R. Grove, FRS, was the Professor of Experimental Philosophy in the Institution.

The Liverpool Royal Institution, modelled on its London counterpart, was founded in 1817 to increase knowledge of science by means of 'academic schools' and public lectures on chemistry and natural philosophy. Lectures on natural and experimental philosophy were also given regularly at the Royal institution of Manchester (1824).

The cause of science was considerably furthered by the establishment of professional scientific institutions such as the Institutions of Civil and Mechanical Engineering. The College for Civil Engineers at Putney quite rightly gave a prominent place to the mathematical and physical sciences. Another very important body in the furtherance of the cause of science was the British Association for the Advancement of Knowledge, which was founded in 1831 and which is still very active.

The Literary and Philosophical Societies were also very effective agents in spreading knowledge of science by means of lectures and discussions. The foundation of the Manchester Literary and Philosophical Society in 1781 was followed by similar societies in Edinburgh, Newcastle, Birmingham, Bristol, Sheffield, Leeds, Kingston-upon-Hull, Yorkshire and elsewhere.

As typical of the activities of these societies may be mentioned the Newcastle Society, founded 1793. The annual reports refer to numerous lectures on natural and experimental philosophy. Within a few years the Society spent well over £1000 on scientific apparatus. The Society's first 'Professor' of Natural and Experimental Philosophy was the Rev. William Turner, who gave courses on mechanics, hydrostatics, pneumatics, optics, electricity, galvanism, magnetism, and astronomy, up till 1833.

Courses of lectures were also frequently given by visiting lecturers such as Dr Lardner who, in 1836, gave 12 lectures on the Steam Engine, Mr Addams who gave 15 lectures on acoustics, optics and electricity in the same year, and W.G. Armstrong who in 1844 gave a course on hydro-electricity which was so popular that several people were unable to gain admission.

These scientific and literary and philosophical societies, through their libraries, meetings and journals, were an important vehicle for the dissemination of mathematical and scientific knowledge. They were instrumental in spreading the desire for higher education which resulted in the establishment of provincial universities.

The Royal Institution: the early years

The Royal Institution was founded in 1799 by the American Benjamin Thompson, better known as Count Rumford (1753–1814), 'for diffusing the knowledge and facilitating the general introduction of useful mechanical inventions and improvements, and for teaching by courses of philosophical lectures and experiments the application of science to the common purposes of life.' Lord Rumford occasionally lectured on heat. It was he who showed how mechanical work in the form of boring a cannon could generate so much heat as to boil a tumbler of water.

By 1801 the Institution had professors of natural philosophy, of mechanics and of chemistry; daily lectures were given and a spacious laboratory and workshop were in use. The lectures attracted such large audiences that Rumford could claim that the Institution was 'not only the fashion but the rage'.

The first Professor of Natural Philosophy was Dr Thomas Garnett, who had acquired fame from his lectures on experimental philosophy and chemistry at Liverpool, Manchester, Warrington, Birmingham and at the Andersonian Institution in Glasgow.

Garnett was succeeded by Thomas ('Phenomenon') Young (1773–1829), FRS. Young was a truly remarkable man. His deciphering of the Rosetta Stone laid the foundation for the study of Egyptian hieroglyphic inscriptions. During his two years at the Institution he gave a remarkable course of lectures on physics, including his great discovery of the interference of light. He is commonly regarded today as the originator of the wave theory of light. His course of lectures on natural philosophy was later published in a work of two quarto volumes. In 1803 John Dalton (1766–1844) succeeded Young and gave three lectures a week.

The lecturer in chemistry in 1801 was Humphrey Davy (1778–1829). His first lectures were on 'Galvanism', a subject then attracting much attention. Long before his lecture was due to begin the whole of Albermarle Street would be blocked with carriages.

With the departure of Rumford in 1802 plans for the instruction of young mechanics were abandoned, and the Institution became a centre for research and for the spread of scientific knowledge among the more cultured class; it became, in the words of Thomas Carlyle, 'a kind of sublime mechanics', institute for the upper classes'.

In 1813 Davy began a long European tour. The party consisted of Davy himself, Lady Davy, her maid and Faraday as a science and general assistant. England and France were then at war, but such was Davy's reputation that the French Emperor gave special dispensation for passports for the party. On the tour Faraday was given many menial tasks to perform, but he bore these with fortitude since his connection with Davy helped him meet some of the great savants of science in Europe. In Paris Davy (and Faraday) met Ampère, Humboldt, Gay Lussac and other eminent French scientists. In July 1814 they were at Geneva, hosts of Professor de la Rive. De la Rive discovered that Faraday was not merely a personal assistant or valet to Davy, but a scientific assistant at the Royal Institution and one who could discuss scientific matters on equal terms with him. Accordingly Faraday was invited to dine with de la Rive along with the Davys. But the Davys objected. As a compromise Faraday was served dinner – but in a separate room. Despite such occasional embarrassments, the tour was for Faraday, a man of humble origin, an education comparable with the years spent at a university by other men.

As Davy's assistant Faraday was now acquiring a certain celebrity. He was called on to speak at the City Philosophical Society where many years earlier he had attended meetings on the occasional free evening. At the City Philosophical Society the members strove to educate each other by evening discussions and lectures – a 'mutual improvement club'. It was here, in January 1816, that Faraday gained his first taste of lecturing. He lectured on the general properties of matter. Later lectures were 'On Radiant Matter' (heat, light, etc.) He made a study of the art of lecturing, having as models the performances of the lecturers on whom he attended at the Institution. Faraday even took lessons from a teacher of elocution. Faraday learnt the art well; in course of time he was the most popular expounder of science in the country. In a letter to a friend, Faraday set down detailed hints on the art of giving scientific lectures – on delivery, on catering to the nature of the audience, on the types of experiments to be performed, on the length of the lecture (so as not to bore the audience), on the pitfalls to be avoided, etc. Such hints would not come amiss today.

Faraday was much inspired by the work on the Continent of men such as Ampère and Oersted. It slowly emerged from their work that magnetism and electricity were intimately connected. It was on Christmas Day, 1821, that Faraday constructed the first electric motor. Faraday's discovery of liquid chlorine in 1823 aroused the jealousy of Davy, who opposed Faraday's election to the Royal Society. He was eventually elected in 1824.

About 1824 Faraday began to deputize for Professor Brande at some of his lectures and in 1827 gave his first course of public lectures at the Institution. Faraday was now 38 years old. The members liked his pleasant and unassuming manner and they soon found that he had a remarkable gift of exposition. Fluency and animation as a speaker were combined with a remarkable ability to devise experiments to illustrate his points.

The Faraday lectures

In 1825 Faraday was made Director of the Laboratory of the Royal Institution. Faraday invited members of the Institution to come to evening meetings in the Laboratory. From these evening meetings grew the Friday Evening Lectures which still continue. In 1826, out of the 17 evening discourses Faraday gave six. At Christmas he gave a course of lectures on electricity for juveniles. Faraday wrote down 80 experiments for his first lecture! So began the annual Christmas Lectures for young people which survive till this day. Faraday had a marvellous rapport with young children. He was supreme as a lecturer and deviser of effective lecture experiments. Faraday was paid the princely sum of £8 15s per lecture at the Royal Institution.

In 1831 Faraday discovered the phenomenon of electromagnetic induction. By rotating a copper disc between the poles of a magnet he was able to generate electricity. The whole science of electrical engineering stemmed from discoveries such as this. In order to make his ideas clear, Faraday postulated the notion of 'lines of force', for he was no mathematician. It is said that Faraday could never understand an experiment thoroughly until he had performed it himself. It was Faraday who first used the method of delineating the lines of force of a magnet by means of iron filings. Nowhere in Faraday's papers is there any mathematics beyond simple arithmetic. It was another great Englishman, James Clark Maxwell, who, 30 years later, expressed in mathematical form the laws of electromagnetism which Faraday had discovered.

In 1833 Faraday was appointed the first Fullerian Professor of Chemistry at the Institution, a Chair he occupied to the end of his life. His investigations into the chemical effects of an electric current led to his *Laws of Electrolysis* (1834). It was Faraday who introduced such now commonly used words as anode, cathode, electrode, cation, anion. Discovery followed discovery. Audiences crowded to hear him at the Friday Evening Lectures and on one or two occasions Faraday had to reprimand his audience for their undignified scrambling for seats.

In 1842 Faraday gave a series of Christmas Lectures on electricity. At about this time the Secretary of the Royal Institution asked Faraday for his advice about the possibility of running evening lectures for the general public. Faraday replied as follows:

> I see no objection to evening lectures if you can find a fit man to give them. As to popular lectures (which at the same time are to be *respectable* and *sound*), none are more difficult to find. Lectures which *really teach* will never be popular; lectures which are popular will never *really teach*.[3]

Eventually the strain of his epoch-making discoveries took their toll on Faraday's health and, exhausted and ill, he was forced to take a long break from his researches and lecturing.

On his return to the Institution the discoveries continued to flow, including the rotation of the plane of polarization of light in a magnetic field. He continued to attract large audiences, drawn from all circles, to the experimental lectures he gave in the large lecture theatre of the Institution. His only assistant was a faithful old soldier.

In 1848 Faraday gave eight lectures on static electricity at the Royal Institution. During the first lecture he received a severe shock from a Leyden battery by accident. Faraday laughed hysterically. The audience, thinking Faraday was really laughing, joined in the fun. Only with some considerable difficulty was Faraday able to continue the lecture.

The circumstances under which Faraday did his work at the Institution were those of penury. During a great part of the 26 years Faraday served the Royal Institution it was only the lectures he gave that kept the Institution alive. At Christmas time Faraday always added lectures for the benefit of children. For ten years, from 1851 to 1860, he gave the Christmas Lectures for young people himself. In 1855 the Prince Consort attended the Christmas Lectures, accompanied by his two sons, the Prince of Wales (later King Edward VII) and Prince Alfred.

At Christmas 1860 Faraday gave the last of his juvenile Lectures on The History of the Candle. Since the lectures began in 1825 Faraday had lectured at Christmas no less than 19 times, usually on electricity or chemistry. Dr Bence Jones, Secretary of the Royal Institution, writes of Faraday:

> In his juvenile Lectures, his simple words and his beautiful experiments, his quickness and his clearness kept the attention, and fixed his instruction in the mind even of the youngest of his hearers, whilst the most practised teacher would find old experiments shown in a new form, which the genius of Faraday only could have invented, and which his handicraft enabled him to carry out.[4]

Faraday was paid £50 for his juvenile lectures.

In 1857 a deputation waited on Faraday to offer him the Presidency of the Royal Society. But he declined the honour. The Royal Society did not forgive him. It refused to accept any papers he submitted: they were only published with his posthumous papers. Faraday was offered a knighthood. This, too, his modesty forced him to decline, saying 'he must remain plain Michael Faraday to the last.' Faraday died at the age of 76, to the end a very religious man. His wishes to have only his relatives and nearest friends at his funeral and to have a grave of the commonest sort were carried out.

John Tyndall (1820–93)

Faraday was succeeded as Resident Professor of the Royal Institution by John Tyndall, who for the next 20 years devoted himself to the Institution and its work, just as his predecessor had done. Tyndall had been a colleague of Faraday at the institution, having been appointed as Professor of Natural Philosophy there in 1853. The relationship between the two men was very close, like father and son.

John Tyndall was educated at a National School in his native Ireland and at Preston Mechanics' Institution. In 1847 he was appointed teacher of mathematics and engineering, surveying and physics at the recently established Quaker school, Queenwood College. The school had been founded in 1839 by Robert Owen but closed in 1845 for financial reasons. It reopened in 1847 as a mixture of an ordinary junior boys school and a technical college for older boys and young men, teaching a wide range of pure and applied sciences, covering especially topics such as the 'steam engine', 'railways', etc. Also on the staff was Frankland, who was later to succeed Faraday as Fullerian Professor of Chemistry at the Royal Institution. Around the middle of the 19th century the staff at Queenwood College must have been one of the most distinguished in the history of school science teaching.

Tyndall was an excellent teacher and lecturer, illustrating his words by beautiful experimental illustrations. In 1853 the newly appointed professor gave a lecture to the Royal Institution which 'took his audience by storm'.

Following in the footsteps of his more illustrious forebear, Tyndall gave innumerable public lectures on several aspects of physics at the Royal Institution. He possessed quite extraordinary powers of popularizing difficult subjects, and he probably did more than any person of his time to diffuse scientific knowledge by the publication of his essays and lectures. He presented the dominant scientific ideas of the century in non-technical language, intelligible to all. In 1863 Tyndall's best-seller – *Heat Considered as a Mode of Motion* – appeared in popular form.

Tyndall did much to encourage experimental work in the teaching of physics. He did so, not only by writing suitable physics textbooks for the schools, but also by showing how teachers, using some ingenuity and determination, could themselves improvise a lot of the apparatus needed.

In December 1875 Tyndall gave the Christmas Lectures at the Royal Institution on the subject of Electricity. With some amplification these were published in 1876 as *Lessons in Electricity* (113 pages). In the Preface Tyndall wrote:

> I had heard doubts expressed as to the value of Science-teaching in schools, and I had heard objections urged on the score of the expensiveness of apparatus. Both doubts and objections would, I considered, be most practically met by showing what could be done in the way of discipline and instruction, by experimental lessons involving the use of apparatus so simple and inexpensive as to be within everybody's reach ... A Price List of the apparatus suitable for the experiments described in these Lessons will be found at the end of the volume. The prices are very reasonable, but the teacher or lecturer may materially reduce the cost by becoming his own instrument-maker ... [5]

A simple school textbook of physics equivalent in standard to about today's Year 9 was Tyndall's *Natural Philosophy in Easy Lessons: A School Text* (178 pages).

Tyndall's main contribution to science was the study of diffusion of light by large molecules and dust – known as the Tyndall effect. The ultra-microscope depends for its operation on the Tyndall effect. In January 1870 he gave a Friday Evening Lecture on Dust and Disease, at which the Prime Minister was present, and in which Tyndall combated the doctrine of spontaneous generation of life. This brought him much criticism and even ridicule, but time has proved him right.

Tyndall resigned his Chair at the Royal Institution in 1887 because of increasing illness. He died in 1893 – not because of his illness, but because his wife inadvertently gave him chloral for his indigestion instead of milk of magnesia.

A colleague with Tyndall at the Royal Institution was James Dewar, who became Fullerian Professor of Chemistry in 1877 when he had already been elected Jacksonian Professor of Natural Philosophy at Cambridge. Dewar held the two Chairs concurrently to the end of his life. Much of Dewar's experimental work, particularly on the liquefaction of gases, was done in the laboratories of the Institution. He would devote endless time and patience on the preparation of his experiments and in the arrangement of his lectures. He brought to the lecture table the skill of the artist. T. Martin writes that Dewar's 'Friday Evening Lectures, particularly those on low temperature work, were occasions to be remembered'.[6]

As Professor of Natural Philosophy, Tyndall was followed by the third Lord Rayleigh, who had retired from the Cavendish Chair at Cambridge in 1884, and who in 1894 had discovered, jointly with Sir William Ramsay, the inert gas argon.

After the retirement of Lord Rayleigh, Sir J.J. Thomson held the Chair of Natural Philosophy from 1905 to 1920. Thomson, in turn, was succeeded by Lord Rutherford (from 1921 to 1937). Though they carried out their brilliant researches in atomic physics in Cambridge, their Friday Evening Lectures, when they described the progress of their work, were great occasions for the Institution. Even on Saturday afternoons, when most people would have left London, Rutherford would fill the lecture theatre to overflowing with an audience of distinguished men of science and students from the London colleges who seized the opportunity to hear the great man.

In 1923 Sir William Bragg, at the time Quain Professor of Physics at the University of London, became Professor of Chemistry and Resident Professor. Bragg's early work had been mathematical. He had been Third Wrangler in the Mathematical Tripos at Cambridge in 1884. Shortly after graduating he was appointed to the Chair of Mathematics and Physics at the University of Adelaide. He knew little physics and so read up the subject on the voyage out to Australia. In 1909 he returned to England as Professor of Physics at Leeds. In 1915 Bragg and his son, Lawrence, were jointly awarded the Nobel Prize in Physics for their work on X-ray crystallography. In 1953 Sir Lawrence Bragg followed in his father's footsteps as Resident Professor and Director of the Davy Faraday Laboratory at the Royal Institution.

Chapter Six

Physics in Elementary Education in the First Half of the 19th Century

Up to the beginning of the 19th century there was no organized system of elementary education in Britain. Adam Smith had proposed in his *Wealth of Nations* (1776) that every parish should support a small school, in which would be taught not only the three Rs but also mechanics. But nothing came of the proposal. Fifty years later J.C. Louden was suggesting a similar plan in his *Parochial Institutions; or an outline plan for a National Education Establishment, suitable to the children of all ranks, from infancy to age of puberty*. In every parish there was to be at least one school-house, with a properly qualified teacher, and containing books and philosophical instruments for instruction in natural philosophy. Thomas Wyse, a Member of Parliament who was very active in movements for the reform of education, in his *Education Reform* (1836) would have had the rudiments of mechanics, chemistry and astronomy introduced even into the elementary schools. There was no reason why an elementary textbook limited to 'notions' of the causes and effects of fire, water, earth and air, of light and colours, of heat and cold, of ice, of day and night, of thunderstorms, of electricity and of steam, etc., might not 'with perfect propriety, be placed in the hands of the simple peasant. In the town schools, further developments might be given in reference to their respective trades.'[1]

In 1811 the National Society was founded for promoting the education of the poor and three years later the British and Foreign School Society was established. These two societies did much to give the elements of education to children who, otherwise, would have gone without it. In the second quarter of the century some of these 'National' and 'British' schools began to introduce elementary instruction in mechanics and science. Alfred Bourne, Secretary of the British and Foreign School Society stated:

> The Society has for many years given prominence in the instruction imparted at its schools to the knowledge of common things, upon which science is based. The Lesson Books adopted by the Society have lessons on such subjects as ventilation, physical geography, the mechanical powers, the atmosphere, the equilibrium of fluids, the velocity of sound, the solar system, the crust of the globe, earths and fossils, and various lessons of the same kind. [2]

From about 1843 Wesleyan Schools were founded in great numbers and Matthew Arnold, an Inspector of Schools, reported that many of these schools also gave instruction in elementary science.

William Lovett (1800–77), the Chartist, aimed to transform society through education. His scheme of education addressed to the working classes in 1837 proposed that in the 'High School' (for pupils over ten years of age) instruction should include the 'physical and natural phenomena of the globe the pupils inhabit, the principles of chemistry and its general application to the arts, trades and pursuits they might hereafter be engaged in.[3]

Lovett, who was self-taught, acquired his enthusiasm for science after hearing Dr Birkbeck lecture at the London Mechanics' Institution. Lovett himself wrote several elementary textbooks on science topics for use in schools. In 1848 Lovett opened the National Hall School in Holborn, in which 'mechanics and the several branches of science' were taught.

Lovett had been greatly impressed by the Birkbeck Schools which were founded in London by William Ellis. These schools were the outcome of a movement parallel to that of the Mechanics' Institutions to educate the children of the middle and working glasses. The first such school was opened in Chancery Lane, Holborn, in July 1848, the subjects taught including 'the elements of mechanics and the physical sciences'. A government inspector reported on Ellis's school that 'the subjects were practically useful, and the teaching was of a high order, displaying energy of mind and extensive knowledge in every department'.[4]

George Combe (1788–1858) was a staunch advocate of introducing elementary science into even the humblest of schools, both for boys and for girls. He would train children to a 'knowledge of things and relations', 'to do more than to know'. In November 1848 he issued the prospectus for a new 'Secular School' to be opened in Edinburgh for children of the working classes. This school came to be called the Williams Secular School, after the Headmaster W. Mattieu Williams. (In 1853 Williams became lecturer in Chemistry and Experimental Physics at the Birmingham and Midland Institute.) In this Secular School the elements of physics and chemistry were taught. An examination of the School took place on 27 July, 1849, with George Combe and James Simpson superintending the proceedings. The following report is given at some length since it indicates the methods used to teach simple physics in an elementary school.

Mr Williams made a drop of water form itself on the end of a glass rod, and fall to the ground. The boys were then questioned on the law of cohesion, which made it round; on the law of gravitation which made it fall; on the nature of fluidity which made it break when it touched the ground; and from these premises they ascended to the laws of the Planetary System, explained how the planets were formed, how the planet Neptune was discovered and how the phenomena of the double stars are accounted for. They deduced from these facts concerning the power, wisdom and omnipresence of God. The laws of heat and evaporation were next expounded by the boys; ... [5]

Combe later reported that 'the lessons on science have been followed with eagerness by the pupils'.

The report stated that when one of the more advanced classes were minutely examined upon the elementary principles of mechanics, the boys 'explained, which much clearness, in answer to questions that were put to them, the different properties of the lever, the wedge, and the screw'.[6]

At the 1850 examination of the school an experiment on the expansion of air due to heat was carried out, and it was reported that the pupils had gone through an elementary course on general properties of matter, on mechanics, hydrostatics, pneumatics, heat and light, and were just commencing chemistry and electricity. The textbooks used were Nichol's *Introduction to Natural Philosophy* and his *Introduction to the Sciences*; also Chambers' *Mechanics* and Tait's *Mechanics*.

James Simpson had advocated very similar views to Combe in his lectures all over the country and in his Philosophy of Education (1834). There was no reason why lessons, illustrated by experiments, on simple statics, dynamics, mechanics, hydrostatics, hydraulics, pneumatics, acoustics, optics, electricity, galvanism, magnetism and on the steam engine, should not be part of the normal education of 12 to 14 year olds.

During the early 1850s several secular schools, on the principles of the Birkbeck and Combe secular schools, embracing a broad range of physical science, were established in various parts of the country. The great improvement in elementary education that resulted from the establishment of these schools is expressed by W.A. Shields, Master of the Peckham Birkbeck Schools.

> That there has been a great improvement effected within the last few years in the education in our common schools is a fact which the experience of all present will enable them to testify ... The schoolmaster is now expected to give such instruction in ... natural philosophy as shall enable him [the pupil] to perceive the principles upon which are constructed, and by which are managed, the various machines in daily use around us.[7]

Richard Dawes

Among the best-known of the elementary schools that introduced simple science into their curricula was that established by Richard Dawes (1793–1867). In the 1840s Richard Dawes, later Dean of Hereford (1850), introduced elementary science into a National School he established in King's Somborne, in Hampshire, making use of simple experiments to illustrate everyday phenomena. In 1817 Dawes was fourth wrangler at Trinity College, Cambridge. He became Fellow and mathematical Tutor, and Bursar of the recently established Downing College. In 1836 he obtained a living at King's Somborne. Dawes was among the new generation of rectors who established village schools with the main objective of raising the moral and religious condition of the pupils.

A distinguishing feature of Dawes's school was the introduction of science. For him, once the mechanical difficulties of reading had been overcome and the basics of simple arithmetic understood there was no kind of knowledge more likely to raise the character of a village school and that of its pupils than simple science. Experience had shown him that the greater the amount of secular knowledge bearing on the lives of the learners, the greater the success. The instruction given was to be related to a culture familiar to the pupils, 'connected with the everyday concerns of common life', drawn from 'things which interest them at the present, as well as those likely to interest them in the future'. From this interest in 'common things' it was a small step to explanations 'of a philosophic kind'. Dawes wrote:

I began giving to the teachers, and more advanced of the school-children, short explanations of a philosophic kind, and in a common-sense sort of way of the things almost daily passing before their eyes, but of the nature of which they had not the slightest conception; such as ... that the air had weight, and how the pressure of this atmosphere helped them to pump up water; enabled them to amuse themselves with squirts and popguns; to suck up water as they called it through a straw; why the kettle top jumped up when the water was boiling on the fire; ... why a glass sometimes breaks when hot water is poured into it, explaining the reason of unequal expansion of the two surfaces, ... these and similar things I found so excessively amusing to them, and at the same time so instructive, that I have scarcely missed a week explaining some principle of this nature, and in questioning them on what had been done before ... In subjects of this kind, and to children, mere explanations, as everyone will perceive, are of no use whatever; but when practically illustrated before their eyes, by experiments, they become not only one of the most pleasing sources of instruction but absolutely one of the most useful.[8]

Dawes's pupils were required to participate in experimental inquiries and practical exercises in an environment which provided opportunities for discussions between teacher and taught. The response to this informal method of teaching science was so effective that within two years Dawes had to build an additional classroom which was, in effect, a rudimentary laboratory. This housed various sectional and glass models of equipment and machinery, an air pump, apparatus for illustrating centrifugal force, chemical apparatus, an electric machine, Leyden jars, a voltaic battery, prisms, lenses, a magic lantern and numerous other pieces of equipment.

Dawes made use of the cheap yet excellent reading books produced by the Irish National Board. These contained a mine of scientific information. *The Fifth Book of Lessons*, which was used by boys in the first class of the school, contained material on astronomy, hydrostatics, pneumatics, optics, electricity and chemistry.

However, teachers with adequate knowledge themselves and with appropriate 'know-how' were in short supply. So Dawes himself gave instruction in simple physics to his teachers at his village school. In 1847 Dawes wrote *Suggestive Hints towards improved Secular Instruction making it bear upon Practical Life*, which was intended for the use of teachers in elementary schools. This book achieved a wide circulation, quickly ran through six editions and became a prescribed text for students in training colleges. After asserting that he thought 'experimental science, a knowledge of common things, ought to form an important part of the instruction in all our training schools' and that 'the great art in teaching children is not in talking only, but in practically illustrating what is taught', Dawes went on to outline the practice at his village school in the teaching of simple mechanics, pneumatics, hydrostatics, sound, heat, optics, astronomy and electricity.

The progress made by his teachers is described by Dawes thus:

The teachers here, who at first knew but little of these matters, are now well qualified to give instruction in them; to teach the mechanical principles of the tools they use, the spade, the axe, the plough; and to explain such things as the common pump, barometer, pair of bellows; metals varying in volume according to the quantity of heat which is in them, why one substance feels colder to the hand than another; ... how water is converted to steam, and again condensed; how their clothes are dried, why they feel cold in sitting in wet clothes; why one body floats in water and another sinks; how much in volume and how much in weight, a floating body displaces of the fluid in which it floats; why, on going to school on a cold morning, they sometimes see a quantity of water on the glass, why it is on the inside and not on the outside, ... All this is mentioned to suggest to the teachers in our schools to qualify themselves for this sort of instruction; and speaking from experience I am quite sure there is no kind of knowledge when intelligently given, more likely to raise the character of the village school than this.[9]

In this pioneering manual of science teaching, Dawes presented his case not by theoretical argument but by a wealth of practical examples within the experience of elementary school children. Experiments using simple apparatus, tables of data, and questions to stimulate discussion and further investigation were included in a long chapter occupying a quarter of the book. D. Layton remarks:

> An infusion of vitality and enthusiasm carries the reader over page after page of fertile and ingenious teaching material. The work is clearly the tested product of an experienced and gifted educator. Here was knowledge at work in a manner which engaged the intelligence of children of the labouring poor, of whom it had previously been taken for granted that they should have 'no business with anything where the mind is concerned'.[10]

The Rev. Henry Moseley, an Inspector of Schools, stated in 1847 that the great excellence of this school, which had 173 boys and girls of average age just over ten years, lay in the instruction it gave in a few, simple principles of natural science, applicable to the things familiar to the children's daily observation. Henry Moseley FRS (1801–1872) was the first schools inspector to be appointed who had a mathematical and scientific background. In 1826 he graduated seventh wrangler from St John's College, Cambridge. He was appointed curate at West Monkton, near Taunton, where he wrote his first book, *A Treatise on Hydrostatics and Hydrodynamics*, in 1830, and in which he made considerable use of the calculus. From 1831 to 1844 he was the first Professor of Natural and Experimental Philosophy, including Astronomy, at the recently established King's College, London.

Moseley laboured unceasingly for science education throughout his life. In 1847 King's Somborne came within his jurisdiction of inspection, and it was at Dawes's school that he found an education based on respect for the culture of the labouring classes. Like Dawes, Moseley believed no branch of secular instruction was likely to be more effective in raising the conditions of the labouring classes than a knowledge of those principles of science which had an application to their welfare and future occupations. For Moseley, too, science was something that could be understood and used without the necessity of actually undergoing the discipline of inquiry. It was legitimate to acquire scientific knowledge from books; the purpose of experiments was to make more vivid and concrete the abstract verbal account. The 'conversational lecture', an oral lesson based on experiments, was an appropriate instructional mode. Pupils could be involved in the experimental demonstrations of the teacher. For Dawes and Moseley any science apparatus required in elementary schools was less for individual practical work by pupils than for experiments in the context of conversational oral lessons.

Moseley was instrumental in setting up the Bristol Trade School in 1856, built as a higher class of school in place of a National School for the sons of tradesmen and the upper class of mechanics. It received some 150 boys a year, aged 12–15, who were taught elementary mathematics, chemistry, physics and mechanical drawing. The fee was £3 a year. In his evidence to the Schools Inquiry Commission (March 1865) Moseley stated that 'there is one principal subject of instruction, round which all the rest are grouped, and that subject is experimental science.'

When asked if he had the apparatus and all the necessary appliances for the instruction and illustration of his lessons in chemistry and experimental physics, Coomber, the Headmaster, told the Commission: 'Yes, every scientific fact that can be illustrated by experiment, and be demonstrated by experiment, is taught by us.' Some of the apparatus was made by the pupils themselves in the school workshop. Mr Coomber reported: 'I do not think there is a lad in my class who studies experimental physics who has not made his own electrical machine.'[11]

The Bristol Trade School and the Navigation School at Hull were conspicuous examples of the successful introduction of science teaching into elementary schools of a more superior kind for artisans.

Elementary Textbooks of Physics

Numerous elementary textbooks of science were published for the use of children during the first half of the 19th century. From these anyone with basic literacy and a degree of application could acquire some knowledge of science. The material was descriptive and non-quantitative.

Many of these early elementary textbooks adopted the 'question and answer' method. The question-and-answer catechisms by Pinnock on navigation, on astronomy, on electricity, on mechanics, on hydro-dynamics, on optics and on natural philosophy, were all fairly good little books, illustrated, and written in easy and familiar language adapted to young minds.

At the end of the 18th century a tradition had arisen of writing children's books in the form of a conversation between parent or tutor and the children. In the course of such conversations some knowledge of science was imparted to children at a level consonant with their abilities. The very popular *Conversations on Chemistry* (1806) of Mrs Jane Marcet, who had attended lectures on science at the Royal Institution, covered the principal topics of Heat. These *Conversations* were the precursor of a similar work on *Natural Philosophy* which went through 16 editions.

The young Faraday was among Mrs Marcet's grateful readers. In 1858, when he was at the height of his fame, Faraday referred to her as his 'first instructress, one able to convey the truth and principle of those boundless fields of knowledge which concern natural things, to the young, untaught, and inquiring mind.'[12] In 1845 Mrs Marcet wrote to Faraday to say that she was holding back the proofs of her *Conversations on Electricity* so that she might include his views on the identity of heat, light and electricity.[13]

Among the best known of such 'conversation' books was Maria Edgeworth's *Harry and Lucy* (1825). In his *Practical Education* (1798) Maria's father, R.L. Edgeworth, had written that children's 'senses should be exercised in experiments, and these experiments should be simple, distinct, and applicable to some object in which our pupils are immediately interested.' This is what *Harry and Lucy* attempted to do. The breakdown of a domestic pump, the failure of a clock, or the breaking of a lock would be starting points for informal, but nonetheless, valuable lessons in simple science.

With the advent of group instruction, as in the monitorial schools, the opportunity of children obtaining first-hand information and making their own experiments become almost non-existent. Accordingly the lessons tended to result more in descriptions of objects and in reading about phenomena. Object teaching in elementary schools was popularized by the advent of Miss Mayo's *Lessons on Objects* in 1830. More than that, this book established a recognized method for lessons of this type that was adhered to for nearly three-quarters of a century.

Though learning science by merely reading about it is deprecated today, much information about physics was given to the children of those times in the course of reading lessons. The children did, at least, get some knowledge of science by this means, and that was better than no knowledge at all.

W. and R. Chambers edited their *Chambers' Educational Course* from 1837 to 1857. This comprised a series of small, cheap, simple treatises on natural philosophy, mechanics and mechanical forces, hydrostatics, hydraulics and pneumatics, optics, acoustics, electricity and magnetism, meteorology, astronomy and chemistry. These textbooks were very extensively used.

A number of periodicals were also important at this time. One that encouraged the study of physics among young children was the *Juvenile Encyclopaedia* or *Monthly Preceptor* which was commenced in 1800. It contained lectures on 'Natural and Experimental Philosophy, adapted to the capacities of young persons'. It also offered prizes amounting to four guineas to boys not exceeding 16 who were able to solve the mathematics and physics questions posed. Similarly, the magazine *Popular Science and Journal of the Useful Arts* (1836) contained a page devoted 'to showing our young philosopher what a great number of interesting experiments he may perform with no other apparatus than a few of the most common articles, which are everywhere at hand ... directions we shall give him.'

One of the most popular textbook writers of the time was the Rev. Dionysius Lardner (1793–1859), the first Professor of Natural Philosophy at the recently established University College, London. Between 1829 and 1849 he edited 133 volumes of his *Cabinet Cyclopaedia*, at six shillings a volume. The science volumes in the series were designed to provide 'a series of works on popular and practical science, freed from mathematical symbols and technical terms, written in simple and perspicuous language, and illustrated by facts and experiments which are level to the capacity of ordinary minds'. Sir J. Herschel wrote the first treatise in *The Study of Natural Philosophy*; Sir David Brewster wrote the section on 'Optics'; whilst Lardner himself contributed the texts on 'Heat', on 'Mechanics' and on 'Hydrostatics and Pneumatics'. The second edition of *A Rudimentary Treatise on the Steam Engine ... for the use of beginners* was published in 1852. This was an illustrated booklet of 132 pages.

Besides very elementary books on physics, numerous textbooks were also written that treated of physics more formally. James Mitchell wrote several simple physics textbooks for children, the best known being his *Newtonian System of Philosophy ... for the Use of Young Persons.*

In 1807 the Rev. David Blair published *An Easy Grammar of Natural and Experimental Philosophy: for the use of schools.* In the preface Blair pointed out that:

> No science can be taught if the student does not *work* or perform operations in it... such easy and familiar Experiments and Illustrations [have been included] as enable the young student to *work* in each science, and at the same time renders its principles intelligible to the lowest capacity.

The textbook contained material on simple statics, dynamics, hydrostatics, pneumatics, acoustics, optics, astronomy, electricity, galvanism, and magnetism. In each section Blair mentioned several experiments that were to be performed.

In 1826 appeared the Rev. John Barclay's well-illustrated *Elements of Natural and Experimental Philosophy*, in which he gave details of various simple experiments that were to be carried out. This was an excellent textbook covering the entire range of physics. Apart from its numerous fine illustrations, its simple language and its frequent references to everyday phenomena and to the performance of simple experiments, each section ended with numerous searching and interesting questions that were not beyond the capabilities of school children.

Very popular as a school textbook both in England and in America, was J.L. Comstock's *System of Natural Philosophy*, which was published in 1842 and which had reached its 195th edition in 1856. This book, with appropriate questions on each page for the examination of scholars, covered mechanics, hydrostatics, hydraulics, pneumatics, the steam engine, acoustics, music, optics, astronomy, electricity and magnetism.

The translation made by E. West of C.F. Peschel's *Elements of Physics* appeared in 1845. In his preface West, after stating that: 'the study of the physical sciences has become of late almost universal ... instruction in these branches of knowledge now forms part of the education given to most boys at school', went on to add that this textbook had been specially adapted for use in schools. This was a good textbook, well-illustrated, and including a lot of up-to-date knowledge on all branches of physics.

Many of the authors of the physics textbooks mentioned show they were aware of and understood those educational principles on which sound instruction in science must rest. Robert Hunt FRS (1807–87), the Professor of Mechanical Science in the Government School of Mines and author of *Elementary Physics* (1851), in a lecture 'On Familiar Methods of Instruction in Science' given in 1854, wished:

> to impress upon all teachers of the young not to attempt to teach science in all its details, but to excite curiosity, stimulate inquiry and quicken the powers of observation. I would teach great phenomena by the most familiar methods, and if I could not make the playthings of the child subservient to my purpose, I should desire the most simple forms of apparatus as my means of illustration.[14]

It was because 'for a numerous class most of the works on natural philosophy are of too abstruse and technical a character' that Hunt published his *Elementary Physics*. This was an excellent textbook, very well illustrated, written in simple language, and incorporating recent developments in the subjects.

Science Apparatus

In 1851 the Rev. C.H. Bromley, Principal of Cheltenham Teacher Training School, applied to the Committee of Council on Education for a grant towards the purchase of scientific apparatus. The Secretary asked Moseley to draw up a schedule of apparatus which could be recommended for use in teacher-training schools, and to provide information on a suitable course of study and on suitable textbooks. He was also asked to furnish a second, less elaborate, schedule of apparatus for use in elementary schools. Economy and simplicity of apparatus were to be the guidelines for both schedules. Moseley spent a year on a national survey of apparatus suitable for teaching science in elementary schools. He drew up three lists of apparatus, costing £10, £15 and £20 respectively, together with more elaborate sets for training colleges, costing £100, £125 and £150.

There were numerous scientific instrument-makers in existence, especially in and around London. E.G.R. Taylor lists more than 40 names for the period 1835–1840. Scientific diagrams, lecture illustrations, sectional models, and apparatus for experiments in the various branches of physics became increasingly available. In response to Moseley's list the firm of Griffen produced a £20 selection of apparatus for use in elementary schools which included items for teaching hydrostatics, pneumatics, electricity and magnetism, together with apparatus and materials for agricultural chemistry.

In 1844 the Department of Science and Art gave grants to teacher training colleges and to elementary schools for the purchase of science apparatus. The Department also undertook to provide models, diagrams and apparatus for purchase at reduced rates. Among the officially approved apparatus were:

Table 6.1

From Thomas Varty, 31 Strand, London: A set of Mechanical Models	63/–
From Taylor and Walton, Upper Gower Street: Set of 15 large Mechanics Diagrams Atwood's Machine Set of Mechanical Powers Bent Lever Double inclined plane Sectional model of steam engine	 15/– 42/– 46/– 10/– 10/– 42/–
From Society for Promotion of Christian Knowledge: Mechanical Models	25/–, 46/–, 92/–

Several training colleges asked Moseley for advice on the sort of science apparatus they should purchase. Moseley warned of the danger of buying expensive and elaborate apparatus, pointing out that some of the most significant work in science had been done with the simplest of materials.

John Gibson, one of the first two school inspectors in Scotland, made reference in his report for 1842–3 to several schoolmasters who out of their own meagre income had bought apparatus and 'had exercised no small share of mechanical skill and dexterity in the construction of apparatus suited to illustrate the lessons on natural philosophy and chemistry'.

Many schools and establishments made good use of the grant for purchase of science apparatus as is evidenced in the replies given to the Devonshire Commission.

The Training and Supply of Science Teachers

The key to success in introducing science into the elementary school curriculum was to be found not in visual and material aids but in teachers who could and would use them. As the Report of the Select Committee on Scientific Instruction pointed out later (1868) 'a hindrance second only to that of the defective elementary education of the pupils is the *scarcity of science teachers*, and the want of schools for training them' [my emphasis]. Without an adequate knowledge of science on the part of the teacher, lessons on the science of common things would degenerate into barren exercises in 'observation and classification'. In Dawes's estimate not one in a hundred elementary-school teachers at the turn of the century was equipped for the task, whilst Moseley's experience led him to wonder if all the teacher training schools could provide one teacher with adequate knowledge of science.

At the beginning of the 19th century the training and supply of science teachers was a casual, unco-ordinated affair. David Stow (1793–1864) claimed to be the first to introduce lessons on natural science into the training of teachers. He certainly was the first to introduce the system – common today – where student teachers give lessons before their colleagues and so benefit from their criticisms.

In the late 1820s Stow founded the Glasgow Normal (Teaching) Seminary. Instruction in elementary science formed a prominent part of the training. In a model school attached to the seminary Stow introduced oral lessons on natural science applied to everyday life, illustrated by apparatus, models and experiments. Stow advocated that the teacher should picture out to the child, by analogy and familiar illustrations, the uses of the pulley, the screw, the lever, the nature of thunder and lightning, and so on. In an advertisement in the tenth edition of his book *The Training System*, Stow claimed that by 1854 2600 male and female teachers had been trained under his system.

To cater for the demand from practising and intending teachers for special courses in the new subjects such as science and in the use of the latest apparatus, the Rev. R. Watson and C. Todd, among others, ran special courses. Watson, who kept an academy in Oldham (1830) which was well equipped with apparatus, gave lectures and practical demonstrations on teaching. About the same time C. Todd, an engineer of Manchester, was running classes in Mechanics, etc.

In 1835 the Committee of Council on Education proposed to set up a State Training College, but this proposal had to be abandoned because of religious controversy. By 1839 both the National and the British and Foreign School Societies had their own training colleges – St Mark's College, Chelsea, belonging to the former, the Borough Road College, London, to the latter. These training colleges gave a general education to young entrants of 15 or 16 years, and also an intensive few months, training to older students.

In 1839 Dr James Phillips Kay, later Kay-Shuttleworth (1804–1877), was appointed Secretary to the newly created Committee of Council on Education. He was responsible for establishing the pupil-teacher system and for the state training of teachers. Kay took a degree in Medicine at Edinburgh in 1827. In 1835 he was made a Poor Law Commissioner (the Poor Law Act having been passed the year before). In 1839–40 he and a friend established the first training college for teachers as a private venture. This St John's College was started in a room above stables in Battersea. The rapid development of steam machinery at this time prompted Kay to add to the curriculum a course of instruction in mechanics and the laws of heat; pupils were required to master the principles and construction of the various kinds of steam engines, following 'the course of the investigator or inventor'.

Thomas Tate, a first-rate physicist and teacher, was appointed in 1840 to give instruction in physics at St John's College. He was also responsible for a whole series of excellent textbooks covering the entire range of physics. Moseley reported in 1847 that 'in calculations connected with the science of practical mechanics this institution (Battersea) is in advance of any other that I am acquainted with'.

At the request of the directors of one of the railway companies, Tate gave evening classes to its employees, in this way giving the work-force something more valuable to themselves in the evening than, as Kay-Shuttleworth put it, 'stupefying themselves with dram-drinking.[15]

Pupil-Teachers

The '*pupil-teacher*' system of instruction was introduced in 1846. It broke up the 'monitorial' system (where older children were used as monitors to teach those younger), which had been in vogue since it was introduced by Andrew Bell and the Quaker Joseph Lancaster at the beginning of the century. Bright pupils aged 13–18 were apprenticed to headteachers for up to five years. These pupil-teachers – Kay-Shuttleworth's 'pioneers of civilization' – were examined yearly and if they acquitted themselves well their headteacher received a bonus. At the end of their apprenticeship pupil-teachers were eligible to compete for Queen's Scholarships which admitted them for up to three years at a training college. The course of training usually lasted two years.

The demands made by the Queen's Scholarship examinations were not severe. In 1846 it sufficed if candidates could read, write, spell, and if they knew the first four rules of arithmetic and some history and geography. By 1850 they were expected to show a rudimentary knowledge of the usual subjects of the elementary curriculum and to answer questions on the practical duties of a teacher. At the end of this period of training a further examination led to certification which entitled successful candidates to proficiency grants towards their salaries.

The study of elementary physics was encouraged by the ruling of the Committee of Council on Education in 1846 that pupil-teachers, at the end of their third year of apprenticeship, would be examined by an Inspector in the elements of mechanics, among other subjects.

By and large pupil-teachers had to depend largely on their own efforts and get what information they could from books. Many small manuals on the methodology of teaching were published for their benefit and a special periodical – the *Pupil-Teacher* – was devoted to the needs of young aspirants to the teaching profession. The lessons illustrate the insistence on mere facts and verbal accuracy that was for long the tradition of the pseudo-science of the training colleges and schools.

When the pupil-teacher system was introduced in 1846 there were 13 normal (i.e. teacher-training) schools in the country – at Battersea, Caernarvon, Chelsea (St Mark's), Cheltenham, Chester, Chichester, Culham, Durham, Exeter, Highbury, Saltley, Winchester, York. By 1852 their number had grown to 39, providing a significant increase in the output of certificated teachers.

Courses of study

Moseley pointed out the necessity of teachers in training going through a course of instruction in science if they were later to teach the subject satisfactorily. All students were expected to follow a broad foundation course in physical science, one element of which, mechanics, was made a prescribed subject of examination for all students at the end of their first year. At the end of the second year the examination was to include Physical Science, covering hydrostatics, pneumatics, electricity, heat, optics and inorganic chemistry. Some idea of the sort of questions in physics set to training college students in 1856 is given in Appendix 1.

Moseley later sent out a notice that no training college would be allowed to present candidates for the examinations in physical science unless the Inspector reported that the college had sufficient apparatus to provide experimental instruction. To comply with this injunction Wesleyan Training College, Horseferry Road, Westminster, which had been founded in 1851, spent £200 in 1856 on scientific apparatus. Dr J.H. Rigg, the principal, reported to the Devonshire Commission that a lecture room, 30 feet by 25 feet, was fitted up for the reception and use of this apparatus.

> It included a good supply of articles and materials for the teaching of chemistry, and also for the illustration of the chief phenomena and principles of practical mechanics, hydrostatics and hydraulics, pneumatics, optics, and heat, frictional and galvanic electricity, and magnetism.

In his evidence to the Commission the Rev. John Cromwell, Principal of St Mark's Training College, Chelsea, reported that between 1851 and 1861, instruction was given in

> applied mechanics, in chemistry, in hydrostatics and in optics ... A laboratory was established, and a regularly paid and competent teacher was employed. Models of all the ordinary machines were bought and the students were instructed in applied mechanics. We have also the apparatus which would enable a lecturer to give courses on heat, light, and electricity.

He went on to describe the practice teaching his students had to do.

> We require every student to go into the practising schools to teach a class of boys, and also to teach a class in the presence of the rest of the students, and no one, even the utmost stranger, would fail to notice a difference between the way in which a young man who has been a pupil teacher manages a class, and that in which it is managed by another young man who has not been a pupil teacher. The pupil teachers' course, therefore, is extremely valuable as a means of suggesting the right method of class-teaching.

In 1855 HMI Matthew Arnold inspected the Borough Road, Southwark, Training College of the British and Foreign School Society. He reported of its teaching mechanics, astronomy, physics and chemistry.

The reports of HMIs[16] show that often elementary science was taught badly in training colleges, leaving the students inadequately prepared to teach science in the elementary schools.

> The professional training was somewhat unreal; it encouraged the cult of 'talk and chalk'; it was tested by 'show' lessons before the inspector, in which it was generally believed that the best impression was made by choosing a scientific topic and working a number of 'interesting' experiments which the class regarded in the light of conjuring tricks.[17]

Despite exhortations made here and there that teachers should illustrate their lessons by experiments, in effect such experiments were merely 'performances' to verify statements made by the teacher. The teaching was entirely didactic; there was no conception of young students themselves doing any real experimental work. The science teaching of the training colleges was largely a preparation for the subsequent stereotyped object lessons in natural history or on the science of common things. Experimental physics at elementary level of the kind *we* have come to associate with the word 'experimental' would have to wait many years for implementation.

Chapter Seven

Physics in Secondary Education in the First Half of the 19th Century

During the first half of the 19th century secondary education for the nation's children was provided in the following establishments: a) *the public schools*, few in number and providing an almost exclusively classical education for the sons of the rich; b) *the grammar schools*, over 700 in number, mostly endowed and nearly all inefficient as far as science (or anything else) went; c) *private and proprietary schools*, of which there were several thousand, of every grade of merit.

A Survey of Schools and Academies in England

1. The public schools

At the turn of the 19th century there existed in England over 700 endowed grammar schools. The better known of these came to be called public schools. In terms of numbers and social prestige the seven great public schools were Eton (founded in 1440), Winchester (1382), Westminster (1550), Shrewsbury (1580), Harrow (1615), Charterhouse (1611) and Rugby (1567).

Up to the middle of the 19th century very little science was taught in the public schools. The education pursued was basically a lifeless relic of the Renaissance – what Thomas Carlyle called 'gerund grinding'. The public and old-endowed grammar schools were hardly touched by science.

2. The grammar schools

The spectacular advances made in science by the beginning of the 19th century added weight to the increasing demands for the inclusion of science in the education of Britain's schoolchildren. However, apart from significant exceptions here and there, in general the English grammar school did not respond to these demands until the second half of the century. As late as 1868 the Schools Inquiry (Taunton) Commission reported that 'Natural Science, with some exceptions, is practically excluded from the education of the higher classes in England'.

Many factors militated against the inclusion of science in the curricula of English grammar schools. A judgement handed down in 1805 by Lord Eldon prohibited Leeds Grammar School from using its endowments for teaching any subjects other than Latin and Greek. Even arithmetic was forbidden. This discouraged any teaching of science in those endowed grammar schools that might have wished to introduce the subject. The judgement was only set aside in 1840 by the Grammar Schools Act.

Understandably there were numerous complaints at the exclusion and neglect of science in the grammar schools. For example, Augustus de Morgan (1806–71), Professor of Mathematics at the University of London, in his essay 'On the Study of Natural Philosophy', inveighed against the neglect of science in the schools of England.

> The present neglect of natural philosophy and natural history will furnish a curious story for after times. It will be on record, that among the first commercial people in the world, who depended for their political existence on trade and manufacturers, there was not, generally speaking, in the education of their youth, any account of the principles, whether of mechanics or of chemistry, which constituted the distinction of their country.[1]

The *Quarterly Journal of Education* of 1832 asked:

> Is it reasonable, or creditable, or decent, that boys of fifteen years of age and more, should know absolutely nothing of the simplest laws of mechanical philosophy?
>
> Is it reasonable that all this should be sacrificed to the supposed attainment of two dead languages? ... They learn a little Latin very badly, next to no Greek, and are ignorant of everything else besides.
>
> We may, then, fairly claim to have some new experiment tried, as the pure Latin and Greek system has been tried long enough.[2]

Sir Thomas Wyse, MP, complained that:

> The middle class ... find no instruction which can suit their special middle class wants. They are fed with the dry husks of ancient learning, when they should be taking sound and substantial food from the great treasury of modern discovery. The applications of chemical and mechanical science to everyday wants ... are all subjects worth as much labour and inquiry to that great body, as a little Latin learnt in a very imperfect manner, with some scraps of Greek to boot – the usual stinted course of most of our grammar schools.[3]

J. Booth, writing in his *Education and Educational Institutions*, in 1846, complained that 'Greek and Latin, like bleeding and hot water, seem to be the universal medicine, the infallible panacea of our educational Sangrados'.[4]

Whilst it is true that the classics formed almost the entire curriculum of many of England's grammar schools, that is not true of all such schools. As early as 1793 the Royal Grammar School at Newcastle-upon-Tyne was teaching mechanics, hydrostatics, optics, sound and astronomy. The annual report for 1834 on Hackney Grammar School states that when some of the higher classes were examined, they 'displayed a considerable portion of ... mathematical knowledge', including physical science.[5]

In 1853 the Rev. B.H. Kennedy, Headmaster of the Royal Free Grammar School, Shrewsbury, gave evidence to the effect that 'for some years past' he had provided one lecture a week on science (physical geography, mechanics, hydrostatics, chemistry, electricity, astronomy) 'with the aid of maps, diagrams, models and experiments'.[6]

Mumford quotes an advertisement from the *Manchester Courier* of 5 October 1836, in which it is stated that an undetermined sum was to be set apart for the payment of a teacher and the purchase of apparatus to be used in the teaching of natural philosophy at the Manchester Grammar School.[7]

The conflict that prevailed in educational circles during the 19th century, between the die-hard classicists and those insisting on modern studies, is reflected in the struggle that took place in Manchester Grammar School. One group demanded a curriculum based on the old classical tradition, aimed at training the mind of the scholar by courses of Latin and Greek grammar *before* allowing pupils to enter upon a specialized course of training in science, whether at a university or elsewhere. The other group, backed by those of the governing body (who were associated with the Manchester Athenaeum, Mechanics Institutions and the like), pressed for a more practical education as was being given in Mechanics' Institutions, in Schools of Design, and in The Athenaeum, which *began* with a study of modern languages and attendance at lectures and demonstrations in natural science.

The dispute between the two groups was settled by the Court of Chancery in 1848. The Court laid down that modern languages and science would form an important part of the studies at Manchester Grammar School. In order to do this, the school would have to appoint 'a mathematical master at a salary of £160 a year' and also 'one or more lecturers in natural philosophy, and to provide proper instruments and apparatus for the illustration of lectures'. However, only a mathematics master – Mr Boardman – was appointed. The appointment of a science lecturer was postponed.

One factor that played an important part in bringing science into higher secondary education was the establishment of the **London University Matriculation Examination** in 1836. The subjects of the examination included mathematics, natural philosophy – 'applied' mathematics and physics, chemistry (including heat), natural history, and astronomy. The examination for honours included calculus, statics, dynamics, hydrostatics, pneumatics, optics and sound. The London matriculation syllabus 1838 and the honours examination paper in physics for 1839 are given in Appendix 2 (page 204).

In his lecture in October 1838 'On the Introduction of the Natural Sciences in General Education', Henry Malden, the Professor of Greek at University College, London, pointed out that the University required 'from every candidate a considerable knowledge of several branches of natural science ... and not only will this knowledge be required in the final examination for Degrees, but no small portion of it will be demanded in the Matriculation examination'.[8]

As Malden is somewhat apprehensive that children may be given a lot of scientific information at school without acquiring the mental disciplines of science, he gives his views on how the physical sciences should be taught in schools. These views (from a Professor of Greek!) are so excellent that they are quoted here at some length.

> It is absolutely necessary that the pupil should be employed with things themselves, and not merely with the names and descriptions of things; and that he should himself observe, himself compare and arrange, himself follow the processes of induction, and gradually be led to originate them; in short, that he

should in a great measure make his science for himself out of facts ascertained by himself, and not receive it ready made either from the lips of a teacher or from books. This, I am aware, is not the speediest way of imparting knowledge nor will it enable the young student to parade extensive acquirements; but it is the true mental discipline; and it is moreover the only way in which the foundations of true knowledge can be laid.

Science learnt merely from lectures or from books is little more than an exercise of the verbal memory, and is positively hurtful to the mind by accustoming it to be content with faint ideas and vague notions ... Half a dozen experiments made by himself will both give him clearer knowledge, and do more to form in him a scientific habit of mind, than a hundred experiments witnessed in the lecture room.[9]

Several educational institutions, such as public schools, grammar schools and private schools and colleges, became affiliated to the University of London and took its matriculation examination. These schools that submitted candidates for the examination would have had to give their pupils some considerable instruction in science.

The increasing popularity of the matriculation examination is reflected in these figures:

Table 7.1

Year	Number of candidates	Number successful
1838	23	22
1840	77	69
1845	113	104
1850	206	190
Total up to and including 1851	1675	1493

Besides these grammar schools where physics was definitely taught, certain branches of physics, such as mechanics, were taught in several other grammar schools as part of the mathematics (or sometimes extra-curricular) instruction. In addition, there were many schools whose foundation charters permitted instruction in navigation, mathematics and 'such branches of education as the Governors shall from time to time deem necessary'. Presumably physics would have been included in this category.

Gainsborough Grammar School, for example, specifically combined a scientific and commercial education with a classical one. Such schools were compelled to extend their curricula if they wished to continue to function, since many parents wanted more than dead languages for their children's education in the new industrial and commercial world.[10] Carlisle mentions some grammar schools which had no pupils at all through failure to include a more 'modern' curriculum.

From the time of the 1840 Grammar Schools Act, which set aside Lord Eldon's judgement, a more liberal and 'modern' curriculum became possible in the endowed schools. King Edward VI Grammar School, Birmingham, was entirely remodelled so as to put into effect the recommendations of its Governors, made in 1836, to the effect that the Governors should have power to engage lecturers or teachers to give instruction in natural and experimental philosophy, mechanics, chemistry, mineralogy, and so on. The headmaster of Leeds Grammar School, likewise proposed in 1842 the inclusion of the 'principles of Astronomy and Natural Philosophy' into the curriculum.

Nevertheless, in spite of the greater freedom provided by the 1840 Act, science had difficulty in infiltrating into the traditional classical curriculum generally. In the 1850s the Hurstpierpoint timetable was such that those boys who could not sing were obliged to learn physical science! Many endowed grammar schools stuck to the classics: the principal purpose of a grammar school (as confirmed by the Grammar School Act) was to teach Latin and Greek. Moreover these schools failed to take full advantage of the grants made available for science teaching by the Department of Science and Art.

3. Private and proprietary schools and academies

The rise of manufactures and industries during the first half of the 19th century led to the establishment of many private and proprietary schools and academies that had a much broader curriculum than just the three Rs and classics. Dr Arnold, the famous headmaster of Rugby, was the first to praise this breadth of outlook.

> There are many commercial schools, especially in the manufacturing districts, where the course of study far surpasses what is here given, and where the instruction in scientific subjects, in chemistry, and in mechanics, is carried to a high degree of proficiency.[11]

E.L. Greenberg has collected evidence to show that out of 288 academies about which definite information is available as to their curricula, 50 offered one or more of the subjects natural philosophy, mechanics and chemistry, in addition to mathematics and geography, which were basic subjects.[12] The ages of pupils that attended these establishments, which catered for the middle and lower classes, varied between five and 20, but the average age of schooling was from 10 to 15.

Schools and Academies in London

Mill Hill School was established by the Congregationalists in 1807. Though organized on public-school lines, it included lectures on natural and experimental philosophy in its curriculum and was one of the schools that prepared candidates for the London matriculation examination.

At the City Commercial School, in Lombard Street, the elements of natural philosophy were taught as an additional subject. The subject was also taught at William Johnstone's Academy in Stanmore, Middlesex. Johnstone pressed for more science teaching in schools. He argued that no education could merit the title of 'liberal' unless it included 'the elements of a greater proportion of scientific proficiency'.[13] He went on to add that 'the advantages which will accrue to education from the use of experimental illustrations of scientific instruments and Philosophical Apparatus, need scarcely be mentioned'.

Between 1814 and 1820 John Herapeth (1760–1868) kept a Mathematical Academy at Knowle Hill, Bristol. In 1820 he moved to Cranford, Middlesex, where he opened an academy 'for the instruction of a limited number of young gentlemen in mathematics, natural philosophy and classics'. His course on natural philosophy included not only astronomy, hydrostatics, theory of tides, mechanics, optics, pneumatics, electricity, magnetism, but also his own researches into the causes of heat, light and gravitation.

In 1827 the brothers Matthew Davenport Hill, Rowland Hill and Frederick Hill opened their school at Bruce Castle, Tottenham. In his 'Plans for the Government and Liberal Instruction of Boys' (1822), Matthew Hill mentions that his upper classes were engaged on scientific exercises such as determining the age of the moon at any given time and determining the space through which a body falls in a given time.

From the *Sketch of the System of Education* (1833) at Bruce Castle, we learn that lectures were given from time to time on various branches of natural philosophy, including chemistry and electricity. The *Sketch* mentions that an attempt had been made some few years previously to introduce the physical sciences into the regular curriculum.

> However, not being fortunate enough to obtain the suffrages of our friends to the change, we were obliged, much against our own inclination to abandon the plan ... [However] having been urged to resume the attempt, we have engaged the occasional assistance of a gentleman professionally occupied in the cultivation and teaching of physical sciences. His instruction, however, is confined to such of our pupils as are directed by their parents to join his class.[14]

Both University College, London, and King's College, London, established schools around this time. In 1832 the University of London School was opened in the building of University College. Pupils were taken up to the age of 16. The teaching of applied mathematics, which included several branches of physics, including mechanics, hydrostatics, optics and sound, was started at University College School in 1835–6. Natural philosophy and chemistry were also taught. At first natural philosophy was an alternative to Greek and these classes of boys who did not learn Greek were 'undoubtedly experimental'.[15] Among the textbooks used were Whewell's *Mechanics, Mechanical Euclid*, and Snowballs' *Mechanics*. By 1840 natural philosophy was no longer a mere alternative to Greek but 'became an integral, though not compulsory, part of the curriculum'.

In 1844 there were 465 boys in King's College School, where physics was taught. In the same year a course of chemistry lectures by Professor Daniell was added to the usual branches of instruction.

In 1845 Charles Tomlinson (1808–97), who had studied science at the London Mechanics' Institution under Birkbeck, was the lecturer on experimental science at the school. Tomlinson was a prolific writer of textbooks of physics. In 1838 he published a *Student's Manual of Natural Philosophy*, an illustrated textbook which contained 610 pages covering the subjects of heat, pneumatics, sound and acoustics in detail to a fairly high standard.

Tomlinson's *Introduction to the Study of Natural Philosophy, for the Use of Beginners* appeared in 1848, costing 10 pence. This illustrated little textbook contained a lot of useful information on simple mechanics, hydrostatics, pneumatics, sound, heat, light, magnetism and electricity. In the same year appeared his *Pneumatics for the Use of Beginners*, also costing 10 pence, and in 1849 his *Rudimentary Mechanics for the Use of Beginners*, priced at one shilling. The latter was an extremely good little book, covering the subjects of statics, dynamics and hydrostatics.

All Souls' and Marylebone District Grammar School, founded in 1832 in union with King's College, was also teaching the elements of physical science in 1849, namely mechanics, hydrostatics and pneumatics.

The City of London School, which was established in 1834, was possibly the first secondary school to set aside a room (in this case part of a subterranean passage which led to a lecture theatre) specially adapted for practical science teaching, in 1847. The elements of natural philosophy were taught, whilst lectures were also given on 'chemistry and other branches of experimental philosophy'. As there were no facilities at first for experimental work by the boys, they were encouraged to carry out simple manipulations at home. The science lectures were given by Thomas Hall. Those pupils who distinguished themselves in the elementary course were formed into 'superior classes' in which instruction, at no further charge, was given in 'the higher branches of Mathematical Science, and the application of it to the Study of Physics'.[16] This was another school that presented candidates for the London matriculation examination.

The Classical, Commercial, Mathematical and Scientific Academy at 38, Kensington Lane, Lambeth, was conducted during the 1840s by Anthony Nesbit (1778–1859), John Collis Nesbit (1818–62) and E.P. Nesbit. The prospectus issued by the Nesbits in 1846 gives a good idea of the actual physics instruction in the Academy. Particular attention was paid 'to the teaching of Navigation, and to the Course of Mathematics and Natural Philosophy required by candidates for the Royal Artillery, or the Honourable East India Company's Service'. The course included the subjects of mechanics, hydraulics, hydrostatics, dynamics and other topics of physics, the application of steam, and chemistry.

After asserting that 'no establishment, in which the study of natural philosophy does not form an essential part of the system of introduction, can be considered equal to the acquirements demanded of the rising generation', the prospectus goes on to mention

> the delight with which our Pupils, after the daily duties of Reading, Writing, Arithmetic, the Mathematics, and the Study of Languages, fly to their recreations of Natural History, Philosophy, and particularly of Chemistry, the gratification with which they attend Lectures, and even aspire to be Lecturers themselves.[17]

For chemistry (which included galvanism), the students had a laboratory appropriated to their own use, where they carried out practical work. The prospectus mentions that 'the collection of philosophical instruments is similarly extensive and complete, comprising air-pumps, electrical, galvanic and magnetic apparatus ... telescopes, optical instruments, mechanical models ...' [18] The pupils at this Academy ranged in age from 10 to 18 years.

In addition to this, Kensington and other similar academies that catered specially for entry to the Royal Artillery and the East India Company's Service, these Services had their own particular training schools. The Royal Military Academy, Woolwich, was one such; another was the East India Company's College at Haileybury.

Schools and academies outside London

Manchester

The academy at Higher Broughton was opened by the Rev. J. Relly Beard (1800–76) in 1826. The course of education included the 'Elements of Science'. In his 'Statement of the Objects and Studies' pursued in his school in 1840, he recommended that the pupils should be led 'through an extensive survey of the elements of most of the Physical Sciences'. In 1840 Beard obtained a partner to run his Story Knolls Boarding and Day School, at which pupils were prepared for business, for engineering and for matriculation in London University. W.H. Herford, who came to Beard's school after some years at Manchester Grammar School, says that Beard's teaching of literature, geometry and the natural sciences 'were all rich, rich feasts after starvation'.[19]

Bristol

Bristol College was founded in 1830 to afford the youth of Bristol and its surrounds the opportunity of obtaining a first-rate classical and scientific education on very moderate terms. 'The Outline of the Plan of Education to be pursued at Bristol College' (1830) after an assertion that 'experience has proved, that a close application to the exact sciences is the best discipline for the mind', goes on to detail the course:

> ... having become acquainted with the Differential and Integral Calculus, to ... Statics and Dynamics, ... and the first three sections of Newton's *Principia*. Thus far he may advance, during the first and second years: in the third year he will be occupied with the principles of Hydrostatics and Optics, and with the remainder of the first book of the *Principia*, as well as ... Physical Astronomy.[20]

It will be seen that the course was more akin to that of a university than of a school. In fact, the proprietors had aimed to model the course on that obtaining in Trinity College, Cambridge.

The Professor of Mathematics and Natural Philosophy was F.W.H. Jerrard, BA, of Caius College, Cambridge. In his evidence given before the Educational Committee of the House of Commons on 28 and 29 March 1836, Jerrard affirms that much importance was attached to natural philosophy and experimental science at the college and that even pupils in the junior department (aged seven to 14) were instructed in natural philosophy according to the plan of Dr Arnot's *Physics*. Jerrard then goes on to assert that the studies which were pursued with the greatest ardour were the physical sciences. In the higher department of the college natural philosophy was studied to the same extent as at Cambridge. Among the physics textbooks were Poisson's or Whewell's *Mechanics* and Poisson's *Hydrostatics*.

Liverpool

One of the half-dozen or so academies in Liverpool giving instruction in physics was that at Great George Street, where practical mathematics, astronomy, natural philosophy and chemistry were taught and for which purpose a Mr Sorley possessed 'an extensive apparatus'. An advertisement dated 1835 states that at the Classical and Commercial Academy in Liverpool, 'more than ordinary attention is paid to Chemistry and Natural Philosophy, illustrated by experiments and suitable apparatus'.[21]

Yorkshire

In 1830 the course of instruction at Brampton Academy, Rotherham, included 'the most useful and interesting branches of Natural and Experimental Philosophy illustrated by an extensive and elegant apparatus'.

The Wesleyan Proprietary Grammar School, Sheffield, was opened in 1838. After mathematics, the whole range of mechanical philosophy was taught. This embraced mechanics and dynamics, optics, hydrostatics, astronomy, and the *Principia* of Newton. This course was only for those 'who, under other circumstances, would be pursuing them at Cambridge'. The prospectus goes on to state that 'lectures are regularly given on the different branches of natural philosophy; such as chemistry, electricity, plane astronomy etc. The instruction is furnished with expensive and splendid apparatus, for the purpose of experiments.' This was another of the schools that submitted pupils for the London matriculation examination.

At the Classical, Mathematical, Commercial and Philosophical Seminary, Holy Green House, Sheffield (around 1833) J.H. Abrahams delivered

> a course of Philosophical Lectures to his pupils annually, which are elucidated upon as extensive and elegant electrical, mechanical, magnetic, pneumatic, galvanic, hydrostatic, and astronomic apparatus as can be produced in any private establishment in the country.[22]

Other private schools and academies

In the 1820s John Shoveller kept the King's House Academy, near Portsmouth. In his 'Essay on the most Effectual methods of Advancing Youth in Scholastic Education', he argues that:

> It is desirable however that the pupils should enjoy the advantage of a convenient apparatus, by which some of the most interesting experiments may be exhibited, and the laws and phenomena of nature investigated and explained. A small room should be fitted up for their use, which ... should contain a Telescope, Microscope, Electrical Machine, Air Pump, and other curious machinery necessary for the experimenter.[23]

In his 'Outlines of Practical Education: or a brief statement of the course of elementary, mathematical, classical, and philosophical studies pursued by his own pupils' (1828), James Butler, who kept an academy at 1, Soho Hill, Handsworth, Birmingham, gives some excellent advice on how physics should be taught. After pointing out that in the teaching of natural philosophy and chemistry, 'the first and most suitable mode of instruction, is the exhibition of experiments', Butler emphasizes that 'The learner should be allowed also to perform experiments himself, under the superintendence of his teacher; this will familiarise him with the facts, the explanations of which will be the subjects of future study.'[24] The textbooks in use at his academy were the excellent scientific treatises that went to form *The Library of Useful Knowledge*.

Other academies which introduced scientific subjects included the Green Row Academy, Abbey Holme, Cumberland, where not only was natural philosophy taught but 'in the several branches of Natural Philosophy they (the pupils) have the use of appropriate apparatus'.[25] In the 1840s lectures were regularly given to the pupils of Dame Joan Academy, Canterbury, on 'chemistry, electricity, astronomy, etc., illustrated by extensive apparatus'. In 1847 Edward Frankland taught the physics and chemistry at Queenswood Hall Academy, Hampshire. The books read for the annual examination in 1845 of the Birmingham and Edgbaston Proprietary School covered the elements of statics, hydrostatics, optics, and the first three sections of Newton's *Principia*.

Resistance to New Ideas

The foregoing pages show that during the second quarter of the 19th century ample facilities existed for middle-class children to obtain a 'modern' education. Whilst these schools, academies and colleges often varied widely one from another in size, efficiency and in facilities for experimental work in science, they all shared a common aim of giving an education more in keeping with the needs of the new commercial and industrial age.

That their efforts in this direction should have incurred the ire of those hardened traditionalists who buried their heads in the sands of Latin and Greek is not surprising. Vicesimus Knox, the headmaster of Tonbridge School, writing in his *Remarks on the tendency of certain clauses in a Bill ... to degrade Grammar Schools* (1821) states that:

> On entering the modern Lyceums, we see an imposing apparatus; ... microscopes, telescopes, electric machines, air pumps, and the instruments and requisites for lectures in chemistry and *experimental philosophy* [my emphasis].

Knox argues, however, that these pieces of apparatus

> to boys of twelve or thirteen, are indeed pretty playthings: but, it is to be feared, that the sciences are above the comprehension of children and that this mode of education, to the exclusion of the classical, is ultimately deceptive.

The Rev. Dr Booth has even more scathing things to say about the private academies.

> ...there is scarcely a school of any pretension in the kingdom which has not its pair of globes, celestial and terrestrial, its barometer in many, an electrical apparatus and sometimes even a chemical laboratory may be found, ... but does not everybody know that such educational apparatus are kept for show,

as part of the advertising economy of the school? How rarely is an attempt even made to give real instruction in the sciences to which those instruments belong![26]

Whilst it is probably true that some of the teachers did not always live up to their advertisements, nevertheless, these were offset by those who got on with the task of providing scientific instruction to the many children whose parents were prepared to pay for such an education.

Schools and Academies in Scotland

During the first half of the 19th century there were numerous schools and academies, both secondary and elementary, that included physics in their curricula. In fact in the 1850 General Report on the schools of the Established Church, Dr E. Woodford commented that 'the subjects of physical science receive so much attention that, in some schools, common reading and spelling appear to suffer by it.' Henry Cockburn expressed the changed spirit in education by his remark that 'the tide is setting strongly against classical education.'[27]

In 1811 the Rector of Perth Academy, Adam Ferguson, was teaching dynamics, mechanics, hydrostatics, hydraulics, aerostatics, pneumatics, acoustics, electricity, galvanism, magnetism, optics, chemistry and astronomy. In 1830 Anderson was still teaching this entire range of physics. That Anderson adopted an experimental approach in his teaching is evidenced by the fact that as early as 1809 he persuaded the Town Council to spend £240 18s 8d on scientific apparatus. M. Russel, in the appendix to his *View of the System of Education at present pursued in the Schools and Universities of Scotland* (1813), stated that the apparatus was extensive, embracing all the more common physics instruments, including the powerful air pump with which Leslie succeeded in freezing mercury by the evaporation of ice. The Rector also gave popular lectures on science and the income thus obtained was used in the purchase of other pieces of apparatus such as a sextant, an electrical machine, a galvanic battery and a 'mountain barometer'.

In 1839 the Town Council made a grant of £30 'to put in proper condition the apparatus for illustrating the theory in the various departments of Electricity and Astronomy'.

Anderson was succeeded at Perth Academy by Thomas Miller. In 1842 Miller requested an annual grant for the upkeep of apparatus, pointing out that he had to buy an electromagnetic apparatus for £3 6s 0d, a gold leaf electrometer for 7s 6d, two globes for 10s 6d, a high-pressure steam engine for £3, and a model of a locomotive engine for £3.

As early as 1804 Ayr Academy bought additional apparatus, costing £507 16s 10d, for instruction in physics, astronomy and navigation. At Elgin Academy 'a course of lectures on Natural Philosophy' was occasionally given, 'illustrated by a neat experimental apparatus, partly public and partly the property of the mathematical master'.

The Scottish Naval and Military Academy, which was founded in 1825, gave instruction in physics. George Lees illustrated his lectures by means of experiments.

Lees was the author of a simple but very good and well-illustrated textbook called *A Catechism of Natural Philosophy*. This was in the usual catechetical form, consisting of a series of questions and answers on statics, dynamics, mechanical powers, hydrostatics, hydraulics, pneumatics, the steam engine and optics. Lees also gave lectures in physics to the Scottish Institution for the Education of Young Ladies (founded in 1834), where the education included 'such a liberal course of instruction in the Physical Sciences as had not previously been attempted in schools'.

In the 'Address to the Inhabitants of Dundee' (1834) we are informed that the Rev. Mr McLaren of the Dundee Academy was teaching:

First session:	Mathematics.
Second session:	As full a course as possible of mechanics, including statics and dynamics – with their application to machinery, etc. – hydrostatics, pneumatics and optics. These subjects to be followed up by an outline of the principal facts in magnetism, electricity, galvanism, astronomy and chemistry.
Third session:	Astronomy, electromagnetism and physical optics, chemistry.[28]

The course on natural and mechanical philosophy at Dundee High School in about 1850 covered statics, dynamics, gunnery, central forces, hydrostatics and pneumatics.

In 1834 chemistry was introduced into the Glasgow High School by Hugo Reid, the master in charge of the Philosophical (i.e. Physics) Department. In 1838 Reid published *The Steam Engine, with the laws of heat and pneumatics*. In the next year appeared his *Catechism of Heat*. In the Preface Reid informs us that:

> The subject of heat was formerly considered merely a subordinate branch of Chemistry. It has now become so much extended, and so many beautiful laws by which its phenomena are regulated have been discovered that it has attained the status of an independent science.

This illustrated textbook, in the form of questions and answers, covered the following topics:

- Effects of heat – expansion, liquefaction; vaporization;
- Conduction, convection, radiation;
- Specific heat, latent heat;
- Sources of heat and cold;
- Nature of heat.

Two years later Reid published his *Natural Philosophy; adapted for self-instruction and use in schools*. Natural philosophy was included in the mathematics course in 1837–8.

In 1840 a Mr W.M. Buchanan was teaching natural philosophy, including acoustics, optics, electricity and magnetism, the textbook used being Wood's

Mechanics and Bridge's *Natural Philosophy*.

In the Proceedings at the Annual Distribution of Prizes (1840) we are informed that in the Natural Philosophy Department:

> The mode of teaching is strictly inductive. The subject of the lesson is illustrated at every step by appropriate experiment and reference to fundamental knowledge. The pupil is taught to reason on the phenomena before him, and to draw his conclusions from data evolved under his own observation, and in many cases by his own manipulations. He is led to acquire a knowledge of facts and principles adapted to his capacity and progress, and being accustomed to take nothing for granted that should be proved, he learns the art of analysing evidence, and lays a solid foundation for future acquirements.

Mechanics, hydrostatics, astronomy and natural philosophy were among the subjects of instruction at Glasgow Western Academy, which opened in 1842. Whewell's *Mechanics* and Bland's *Hydrostatics* were among the textbooks in use. When it was proposed to give lectures on Natural Philosophy to the pupils of the Edinburgh High School every Saturday, James Pillans, the Rector, in his 'Three Lectures on the Proper Objects and Methods of Education ... and on the relative utility of Classical Instruction' (1836) argued against 'the new-fangled notions and ill-concerted projects.' He maintained that lectures attractively delivered and illustrated by showy experiments, were 'calculated to captivate the senses than inform or discipline the understanding' and would render 'more and more unpopular among the boys the mental discipline of laborious tasks'.

However, lectures were introduced in 1848. Whilst the first and second classes studied 'Introduction to the Sciences', the third and fourth classes studied mechanics, astronomy being left for the fifth class.

Physics for Girls

Secondary education for girls was provided either at home by parents or governesses or in private schools. Mrs Margaret Bryan, a physicist in her own right, maintained such a school in the early years of the century at Bryan House, Blackheath. In 1806 she published her *Lectures on Natural Philosophy*, 'originally written, and principally intended, for the instruction and improvement of my pupils in the knowledge of physics'. Mrs Bryan's very comprehensive and well-illustrated textbook of 388 pages is full of the most up-to-date knowledge of the simpler topics of physics, and her repeated references to the experimental method is admirable. The book was much used.

Another who taught physics to her pupils was Mary Carpenter (1807–77), who ran a girls' school in Bristol well into the 1840s. She obtained her training in science from her father, Dr Lant Carpenter.

In 1831 Mary Somerville (1780–1872) published her *Mechanism of the Heavens* which established her reputation. She had been educated near Edinburgh at a fashionable boarding school for girls where she worked hard at Latin so that she could read Newton's *Principia*. In 1834 appeared her advanced and up-to-date textbook *On the Connexion of the Physical Sciences* in order 'to make the laws by which the material world is governed more familiar to my country women'. This book reached its eighth edition in 1849.

Chapter Eight

Physics in Elementary Education in the Second Half of the 19th Century
T.H. Huxley (1825–95)

By the mid-1850s many of the conditions necessary for the introduction of science into the elementary school curriculum had been met: books had been published, apparatus was being produced, the supply of trained teachers was growing, and an administrative structure was in place to support and encourage the development. But much remained to be done, as is shown from the evidence given to the Royal (Devonshire) Commission on Scientific Instruction and the Advancement of Sciences (1871–5) by the Rev. Fred Watkins, who was an inspector of elementary schools (Yorkshire) under the Department of the Privy Council for 27 years.

Question: Has any instruction in elementary science been given in any of the schools which have come under your observation?
Reply: Hardly any regular instruction.
Question: To what extent has science been taught in any of the day schools?
Reply: In the day schools with which I have to do it is hardly taught at all. It is only taught in the reading lesson in the elementary books on certain scientific subjects; but these lessons are merely read over. There is no direct and regular course of instruction given. I do not think I can name a single day elementary school where there is any direct and continuous course of instruction.

In his Census Report in 1851 Horace Mann pointed out that out of a child population of five million in the country, well over one million children never attended school at all. And for those who did the average length of school attendance was between one and two years. (It was not till 1876 that education was made compulsory throughout the country.) Many of those who did attend school probably fared worse than those who did not. Matthew Arnold reported on 'the unawakened and uninformed minds of schoolchildren – even of those passing examinations; passively repeating and listening'. He adds: 'In four years I have never known a child to ask the teacher a question.' Large classes and a brutal kind of discipline were the order of the day.

The year 1854 was a significant one for the introduction of science into the curriculum of English schools. In that year an Educational Exhibition was held, during which a series of lectures was given. Among the speakers were some of the most eminent scientists of the day. Michael Faraday, in a lecture on 'The Education of the Judgement', stressed the importance of cultivating a scientific spirit; John

Tyndall, Professor of Natural Philosophy at the Royal Institution, lectured on 'The Importance of the Study of Physics as a Branch of Education'; and Professor T.H. Huxley delivered a famous address on 'The Educational Value of the Natural History Sciences'. Huxley became the giant of the protagonists for science teaching. Whilst Huxley's was not the only voice crying for more science to be introduced in the schools, his great influence as a scientist and his marvellous powers of exposition helped him to do more than perhaps anyone else in the century to further the cause of science education.

Thomas Henry Huxley was born in 1825, in Ealing, Middlesex, the seventh of eight children. In 1833 he attended Ealing School where his father taught mathematics. But this appointment ended two years later, and so did Huxley's scholarship at Ealing. Eventually he won a scholarship to Charing Cross Medical School. When he was 21 he joined *HMS Rattlesnake* as an assistant surgeon and spent the next four years sailing the southern seas. When he returned his natural history researches brought him some considerable renown.

In 1854 Huxley was appointed to a Lectureship in Natural History at the Government School of Mines in Jermyn Street, London. Huxley was instrumental in transforming the school into the first Normal (Teacher Training) School of Science in 1881 and then into the great Royal College of Science.

In Huxley's opinion *all* children should be taught science. This was to be done through 'physiography' (*erdkunde*), physical geography or 'a general knowledge of the earth'. Huxley himself was the author of a little textbook on physiography which had a wide sale and made the subject very popular for many years. Following on a grounding of physiography the child, in his view, should be taught some physical science. But the physical sciences must be taught practically.

> If scientific instruction is to be dealt with as mere bookwork it will be better not to attempt it, but to stick to the Latin Grammar which makes no pretence to be anything but bookwork ...
>
> If the great benefits of scientific training are sought, it is essential that such training should be real – that is to say that the mind of the scholar should be brought into direct relation with fact; that he should not merely be told a thing, but made to see, by the use of his own intellect and ability, that the thing is *so*, and not otherwise ...
>
> Do not be satisfied with telling him that a magnet attracts iron. Let him see that it does; let him feel the pull of the one upon the other for himself ... Pursue this discipline carefully and conscientiously, and you may make sure that, however scanty may be the measure of information which you have poured into the boy's mind, you have created an intellectual habit of priceless value in practical life.

Huxley continues:

> In advocating the introduction of physical science as a leading element in education, I by no means refer only to the higher schools. On the contrary, I believe that such a change is even more imperatively called for in those primary schools in which the children of the poor are expected to turn to the best

account the little time they can devote to the acquisition of knowledge. A great step in this direction has already been made by the establishment of Science Classes under the Department of Science and Art; a measure which came into existence unnoticed,but which will, I believe, turn out to be of more importance to the welfare of the people than many political changes over which the noise of battle has rent the air.[1]

The Department of Science and Art (1853–98)

About the middle of the 19th century it was generally recognized that Great Britain was almost a hundred years behind the Continent in the teaching of science. The newly united Germany had its academies of science and its *realschules*. Science had been introduced into the Prussian *gymnasien* in 1816. In France eminent scientists and mathematicians, such as Laplace and Lagrange, were giving instruction in the Normal (i.e. teacher-training) School which had been founded in Paris by the Republic in 1794.

In 1851 a Great Exhibition was held in Hyde Park. Britain fared badly: all the prizes went to the Continent. The poor showing of British products at the Exhibition brought forth angry protests. J.A. Lloyd remarked:

> Men of the highest eminence in science, the arts, and manufactures, have witnessed ... the rapid advances which they (the Continentals) are making in almost every branch of art and taste, and they have rightly attributed these advances to the improved system of scientific practical education which is now so easily obtainable in many cities and provinces on the Continent.
>
> Every native of France has easily within his reach a system of excellent theoretical education in all branches of Arts, Sciences and Manufactures, which is carried on simultaneously with practical illustrations, derived from materials, models, machines and tools of every description and which the museums of their colleges possess for demonstration.

Lloyd went on to propose that

> the balance of available funds derived from the Exhibition of 1851 be devoted to the endowment of a College of Arts, Sciences and Manufactures in London, resembling the Central School of Arts and Manufactures at Paris.[2]

Dr Lyon Playfair argued in his 'Lectures on the Results of the Great Exhibition' that

> Until our schools accept as a living faith that a study of God's works is more fitted to increase the resources of a nation than a study of the amours of Jupiter or Venus, our Industrial Colleges will make no material headway against those on the Continent.[3]

Even Parliament was now ready to grant money for science and technical teaching. In 1853 the Committee of Council on Education offered grants for the purchase of apparatus provided the school set aside a room as a laboratory and someone who had studied experimental science undertook to lecture upon it. Mr. John Griffin was appointed to distribute the apparatus which consisted of items such as:

> Microscope and microscope apparatus £1 5s 3d.
> Magic lantern with astronomical slides £1 5s 0d.
> Electrical machines, Leyden jar, head of hair, galvanic battery, galvanometer for decomposing water; Leslie's cubical tin box (Heat).

In 1852 a Department of Practical Art was established. The following year a Science Division was added and the two Departments merged to form the *Department of Science and Art.* The task of organizing classes for instruction in science was given to Dr Lyon Playfair. However, little was achieved in the first years of the Department's existence. Invitations were sent out to localities to take up the subject of science instruction but only 11 places in the United Kingdom responded.

Since the nation needed a much increased knowledge of science among its people, the Department of Science and Art decided it would supply this need by paying for it. So began the era of 'payment by results' . In a letter to Tyndall (17 January, 1857) Playfair justified 'payment by results'.

> The State wishes a certain knowledge of science; it is willing to pay for this ... If Queenswood (College) or the Royal Institution or a National School at Battersea, or an Artisan at Poplar produces a certain knowledge of Science in its pupils, we will pay those pupils for proving that they have this knowledge and we will go further and pay the Teacher so much for every pupil whom he has sent up with the required standard of knowledge.[4]

It was hoped in this way to build up a cadre of teachers who would be tempted to take up elementary science teaching.

Playfair resigned from the Science and Art Department in 1858. He was succeeded by Lt Donnelly of the Royal Engineers. Donnelly also appreciated that strong incentives had to be offered if science was ever to get into the schools. In 1859 the Department started a system of grants to schools or classes giving instruction in science and also of payment to teachers based on the results of an examination of their pupils. The examinations were intended to test candidates in evening and other classes rather than full-time schools. In his evidence to the Schools Inquiry Commission in 1865, Robert Lowe, Vice-President of the Education Department, spelt out the simplicity of the plan.

> The plan is this: There are examinations for science masters, a man who passes the examination is certified and registered as a science master. Then there are examinations all over the country, on particular days, on papers which are sent down from the central office, of the pupils of those masters. Those pupils pass or not. According as they pass the masters receive certain gratuities for teaching them, it having been previously ascertained that those masters had the requisite

knowledge to teach. The result of that is that 466 thousand of young men are learning science throughout the country, and making great progress and proficiency, without any expense at all to the Government for finding buildings, schools or libraries, or anything else. That result has been obtained by the efficiency of this plan, paying the master for the results obtained and taking security by a previous examination that the master is competent to teach it. That is the most successful thing I have seen.

The Department's plan succeeded; the teaching of the elements of science was given a great impetus. Whilst in 1851 there had been only 38 science classes in the country with 1300 pupils, in 1861, encouraged by the South Kensington (Science and Art Department) grants, there were 70 science schools with 2543 pupils.[5] By 1872 there were 938 science schools with 36 783 pupils.

Huxley was one of the principal examiners for the Science and Art Department. He stated:

It means that by holding out a very moderate inducement, the masters of primary schools in many parts of the country have been led to convert them into little foci of scientific instruction, and that they and their pupils have contributed to find or to make time enough to carry out this object with a very considerable degree of efficiency.[6]

From the syllabuses laid down by the Science and Art Department, we get some idea of the examinations it conducted.

In *Voltaic Electricity* the candidate ought to be able to state precisely how voltaic electricity may be generated; to describe Volta's pile., and his crown of cups; also the batteries of Daniell, Grove and Bunsen ... He must be able to measure the strength of an electric current and he is strongly recommended to master thoroughly the law of Ohm, regarding the mutual relations of electromotive force, resistance, and strength of current. He ought to be acquainted with the so-called polarisation of metallic plates between which a current passes through a liquid and to show how this is avoided in Grove's battery. He ought to be able to give a clear description of some one form of the electric telegraph. He ought to be acquainted with the physiological effects, and with those of light and heat produced by voltaic current; and to show the dependence of the heat on the strength of the current, and on the resistance which it encounters. It would also be well to master as much of the phenomena of induced currents as would enable the candidate to explain the action of the galvanising apparatus used by medical men.[7]

A perusal of the actual examination papers set indicates the change in thinking that has occurred since then, as to the sort of questions that should appear in an examination paper. In the 1860 physics examination paper, 21 questions were set on 'mechanical physics', but many of the questions required little or no reflective thinking. The answers required mere definitions – of momentum, moving force, velocity, accelerating force. One typical question was: *Distinguish between uniform and accelerated motion. Give examples of each. Show that when no force acts the motion must be uniform.* Other questions involved the purely arithmetical working out of examples. For instance: *The specific heat of air is 0.25. How many cubic feet of air will one cubic foot of water warm 1 degree by yielding up to the air a single degree of its temperature?* Several questions required merely descriptions as answers, for example, the working of such things as the hydraulic press, turbine, boiler safety valve.

The Devonshire Commission reported that up to 1861 the system of teaching science in elementary schools was effective when undertaken by well-trained teachers. The Second Report of the Commission states:

> Indeed, we have evidence of highly competent authorities to show that the scientific instruction which was given by ordinary elementary schoolteachers ... before the introduction of the Revised Code in 1862, was in many cases sound and valuable in itself and beneficial to the pupils.[9]

In its report on the state of science teaching in the training colleges the Devonshire Commission stated:

> With respect to the Training Colleges we find instruction in the elements of science was given to students in their second year. The Committee of Council on Education made grants in aid of the stipends of masters specially qualified to teach the elements of science. In some training colleges a sound, though limited, instruction in the elements of some branches of physical science was thus added to the usual instruction in elementary mathematics. In the model schools attached to the training colleges the students were practised in teaching the rudiments of science, chiefly, though not exclusively, in illustrations of appropriate lessons. ... Such instruction enabled skilful teachers to make the lessons the occasion of conveying important explanations of natural phenomena, and some rudimentary acquaintance with the most easily understood laws of nature; although neither in training, nor in model schools attached to them, were the methods of communicating the rudiments of scientific knowledge by means of systematised object lessons always sufficiently cultivated. The scientific instruction received by the teachers in the training colleges was, however, of great value irrespective of such use in the rudimentary lessons; for all such knowledge tends to enlarge the intelligence of the teacher, and enables him to give better and more interesting instruction.[10]

Even though science was not compulsory for the Teachers' Certificate in the 1850s and 1860s, many training colleges, such as Battersea Training College and Wesleyan Training College, Westminster, formed classes to prepare for the Science and Art examinations. Students who possessed Science and Art Certificates could always get an extra remuneration when they became teachers in the day schools. They could, in addition, teach science classes in the evenings and so augment their modest incomes.

To solve the crucial problem of an adequate supply of teachers a strong, central School of Science was proposed. But the proposal did not materialize. In its stead was adopted what Archer calls 'the miserable substitute of an examination for science teachers'. A special examination was instituted for teachers and a certificate (of three grades) was issued recognizing the amount of knowledge obtained. This examination in science for teachers was first held in May 1861. The teacher's salary was augmented by £20, £15, or £10 for each subject taught, provided that ten or more students had passed an examination in that subject, and that each student had received at least 40 lessons from the teacher during the year.

In order to assist teachers preparing for the examination the Science and Art Department arranged a series of lectures by eminent scientists. Tyndall gave the lectures on experimental physics, Huxley on zoology, and Hofmann on chemistry.

The growth of science teaching in elementary schools and teacher training colleges received a set-back in 1861 when Robert Lowe, in an effort to reduce government spending, introduced his Revised Code. Instead of the various grants schools could accrue, school grants were calculated on the basis of pupils' examination performance in only reading, writing and arithmetic. This resulted in a restriction of the range of the elementary school curriculum to the three Rs and religious knowledge and a general curtailment in the syllabus of training colleges. Up to the Revised Code of 1861 inspectors examined all the subjects taught in a school. After 1861 inspectors examined only those subjects which counted for grant, namely the three Rs and religious knowledge. Inevitably many schools and training colleges abandoned special subjects like mathematics and physical science, for which no grants would be forthcoming. Mr J.C. Buckmaster, Organizing Officer of the Science and Art Department, stated in his evidence that 'scientific teaching in training schools (as at Battersea) was very inferior to what it was ten years ago ... The Revised Code put an end to science teaching in training colleges.'[10] (Some schools and training colleges did, of course, continue with science so that their pupils could do the examinations of the Science and Art Department.)

In 1868 payments to science teachers were made only 'for instruction of students of the artisan or weekly wages classes' of £1 for every second-class pass and £2 for every first-class pass in either the elementary or advanced papers, and £4 for a first-class pass in the honours paper. Also, grants could be given for science teaching in elementary schools provided this instruction was outside the regular teaching hours, since grants for this were already provided by the Committee of the Privy Council on Education. This latter rule gave a tremendous boost to evening classes in science throughout the country. It was in the evening schools that many elementary teachers learnt their science as best they could.

In his evidence to the Devonshire Commission (3 July 1868) Huxley was full of praise for the achievements of the Science and Art Department in its attempts to further the cause of science teaching.

> With respect to its success, I confess I think that it is one of the greatest steps ever made in this country towards spreading a knowledge of science among the people. I have long looked upon any spreading of science from above downward as utterly hopeless, and it appears to me that this department has set on foot a system for spreading science from below upwards.

Professor Tyndall, who set the examination papers in acoustics, light and heat, and in magnetism, for the Science and Art Department, was also fulsome in his praise for the effect the examinations were having in furthering knowledge of science. 'I have no hesitation in saying that a vast quantity of perfectly sound knowledge of a most valuable character is sown broadcast throughout the country, through the agency of these examinations.'

In 1869 the Department paid £17015 in grants to teachers, the average payment being £35 a year. Robert Lowe reported in 1865 that 'Some teachers of science make £150 a year, in fact one man made £250 a year, and we were obliged to put some little check upon it. There is a good deal of money to be made out of it.'[11]

In due course teachers came to look on the Science and Art examinations as a means of increasing their emoluments. Teachers would try to pass examinations in as many subjects as possible, thus qualifying them to teach those subjects in their schools, which meant they could augment their salaries still further. Self-taught students were collecting the science certificates of the Science and Art Department by the dozen.

The get-rich-quick teachers used the easiest expedient possible – cramming. One of the witnesses who gave evidence to the Devonshire Commission was Louis Miall, Secretary of the Bradford Philosophical Society. This is what he said:

> I have a number of somewhat serious faults to find with the system as at present carried out. The great objection, in my opinion, is that the regulations of the department do not encourage what I should call a real style of teaching; that the teaching of scientific subjects, which ought, as I imagine, to be of a highly practical character, is very largely conducted by such means as reading out slowly notes to be taken down verbatim, and committed to memory; or, again, by a large use of elementary text books which are made as condensed as possible, and are in some cases almost learnt off by heart by frequent repetition.[12]

Huxley, and others, did not take long to see some flaws in the Science and Art system. Huxley wrote:

> I dare say you have heard of that proceeding, reprobated by all true sportsmen, which is called 'shooting for the pot'. Well, there is such a thing as 'teaching for the pot' – teaching, that is, not that your scholar may know, but that he may count for payment among those who pass the examinations.[13]

In 1870 25 200 candidates took the elementary stage papers, and of these 11 835 failed (a 46.96 per cent failure rate), 4728 took the advanced stage papers and of these 1723 (36.44 per cent) failed. John Tyndall wrote:

> Many teachers evidently set before themselves, as the main object of their efforts, the bettering of their own pecuniary estate; and a good number of them appear to entertain the notion that this end is more likely to be gained by increasing the number of candidates than by aiming at more excellent instruction. Of the 7,000 pupils recently examined in physics, many ought never to have appeared at all. Their incapacity must have been known to their teachers, provided the teachers themselves be not utterly incapable. Some method, I think, ought to be devised which should not only deprive such teachers of any advantages expected to arise from mere numbers, but which should also subject them to a definite penalty for the perfectly fruitless trouble and expense which they entail upon the Department of Science and Art.[14]

There were other complaints about certain aspects of the Science and Art Department examinations. There was no practical examination and this led to cramming. The Science and Art Department examinations revealed a total absence of any practical or laboratory instruction in the science classes from which the examinees came. It seems that in many cases even such simple aids as blackboard and chalk were neglected. Teachers with little or no experience would slavishly cling to the method by which they themselves had received a modicum of often imperfect knowledge.

Huxley appreciated that if science was to be taught as it should, the first requirement was that the teachers of science themselves should be taught how to do so.

> The great blunder that our people make, I think, is attempting to teach from books; our schoolmasters have largely been taught from books and nothing but books; and a great many of them understand nothing but book-teaching. The consequence is that when they attempt to deal with science teaching they make nothing out of it. If you are setting to work to teach a child science you must teach it through its eyes, and its hands and its senses ....[15]

> One of the most important steps that could be taken towards improving the results of the examinations in science, would be to teach the teachers how to give instruction in science. Teaching in England is pretty much a matter of chance, and the mass of the people are ignorant of the fact that there is such a thing as a scientific method in teaching ... [For teachers] training would have to be special; that it to say, it would not do to bring the teachers up merely to hear the lectures which are given in a course of general science. A teaching which would suit students is not precisely what one would require for the teachers ... I should give a special course of a dozen or twenty lectures for the express use of teachers, showing them how to teach.[16]

In his evidence to the Devonshire Commission, Tyndall, too, pressed for more enlightened methods in the training of science teachers.

> It is, therefore, an object to be aimed at, the founding of institutions in which teachers should be instructed in the use of apparatus sufficiently cheap, simple, and effectual, to be introduced with profit into class instruction.
>
> I would here mark that although no scheme of education in physics is even approximately complete without illustrative experiments, an able teacher, even in the absence of apparatus, can do a great amount of good. It is possible, by the judicious use of the blackboard and chalk, and of simple models, to convey clear conceptions of various parts of physics, so clear, indeed, that, should the pupil afterwards witness the experiments, he shall witness that of which he had previously an accurate, though, it may be, an incomplete conception. Indeed, even where apparatus exists, the performance of experiments ought to go hand in hand with this diagrammatic exposition.[17]

The Science and Art Department initiated a course for science teachers in London in 1869 'with a view,' as Huxley said, 'to converting them into scientific missionaries to convert the Christian heathen of these lands to the true faith.' In the first year 26 teachers attended a one-week practical course in chemistry given by Dr Frankland at the Royal College of Chemistry.

The experiment was a great success. Regular courses and also vacation (during June, July and August) courses for science teachers from outside London followed at the Royal School of Mines, Jermyn Street. No fees were charged, all expenses were paid, and a bonus was given for attending. These vacation courses, which lasted six weeks, covered biology, chemistry and physics. The object of the courses was to show easy and inexpensive methods of giving experimental demonstrations to classes. There were so many applications from science teachers to attend these six-week courses, selection was based on the results of the May Science and Art examination.

The course on physics (light) was given by Dr Guthrie. He had 204 students at his course. 'The most important feature was the practical work through which everyone passed.'[18] During the six-week course the teachers were shown how to set up and manipulate simple apparatus such as they should use in their teaching. A set of apparatus was supplied to those teachers who took the course. The instruction given in these courses was extremely valuable, so much so that some of these teachers were sufficiently capable to teach science even in secondary schools.

In 1870 the Science and Art Department awarded grants for laboratory instruction in chemistry. A few years later similar grants were given for *practical* teaching of physics and other subjects. A grant of £1 per student was paid towards laboratory expenses provided the laboratory of the school was fitted up with all the apparatus listed by the Science and Art Department and provided also that each student did at least $37\frac{1}{2}$ hours of laboratory work.

In 1871 Officers of the Royal Engineers were appointed as District Inspectors for the Science and Art Department, supervising the examinations and inspecting schools, during which 'the teacher might be called upon by the inspector to show his ability to use any apparatus'. The Royal Engineers were chosen for this important role because 'you could hardly find a numerous corps of scientific inspectors, at present, except in that particular body.' In 1879 39 officers of the Royal Engineers were used in school inspection and 58, ranging in rank from Major-General to Lieutenant, in supervising the examinations. The practice of employing Royal Engineers as inspectors continued up to the 1890s, when it finally became possible to dispense with their services.

In addition to courses for science teachers the Government School of Mines provided lectures in science for working men, who attended in extraordinarily large numbers. In due course the School of Mines became a teacher training school 'primarily intended for the instruction of teachers in the various branches of Physical Science, and of students of the industrial classes selected by competition in examinations of the Department of Science and Art.'

Despite their defects, the examinations of the Science and Art Department did play a significant role in bringing science into the schools. Huxley regarded them as 'the most important engine for forcing science into ordinary education'.

> Without presuming in any way to speak in the name of the Department, I think I may say, as a matter which has come under my own observation, that it is doing its best to meet all these difficulties. It systematically promotes practical instruction in the classes; it affords facilities to teachers who desire to learn their business thoroughly; and it is always ready to aid in the suppression of pot-teaching.
>
> All this is, as you may imagine, highly satisfactory to me. I see that spread of scientific education, about which I have so often permitted myself to worry the public, become, for all practical purposes, an accomplished fact.[19]

The School Board Era (1870–1902)

Forster's 1870 Education Act established a state system of publicly elected elementary education boards. Numerous School Boards were set up throughout the country. Many of these drew up schemes of education which included the teaching of science in elementary schools. The most powerful of these, such as the London School Board, had a great influence on the curriculum of the elementary schools. In association with the Science and Art Department they did much to spread the teaching of elementary science. Huxley was the moving spirit behind the London School Board. In 1870 a syllabus Committee was set up, with Huxley as Chairman, 'to consider the scheme of education to be adopted in the Public Elementary Schools and to report thereon to the Board'. Having taken evidence of 'able and experienced teachers with a view to ascertaining the amount and quality of the instruction which has hitherto been found practicable to teach to children' the Committee produced a syllabus remarkably like the syllabus of today in elementary schools: morality and religion, the three Rs, grammar, elementary science, geography, history, elementary social history, music, drawing, physical education.

For Huxley the foundation of science teaching was systematized object lessons. As he stated later:

> The man who has a real knowledge of science can make the commonest object in the world subservient to an introduction to the principles and greater truths of natural knowledge. it is in this way that your science must be taught if it is to be of real service. Do not suppose any amount of bookwork, any repetition by rote of catechisms and other abominations of that kind are of value for our object. That is mere wasting of time. But take the commonest object and lead the child from that foundation to such truths of higher order as may be within his grasp ... but in the respect to the scientific object lessons you want teachers trained in a manner different from that which now prevails.[20]

Such lessons in elementary physical science, to extend over six years, were to serve as an introduction for the work leading to the examinations of the Science and Art Department.

By today's standards and practices much of this science teaching could be criticized. It consisted largely of oral lessons, and demonstrations by an instructor. One headmaster considered that in many cases object lessons were merely clever performances by the teacher, requiring no mental exercise by pupils. The 1875 Code condemned as worthless merely verbal lessons in those subjects such as science which required experiments and illustrations. But when one considers that the classes of those days numbered 60 to 70 children, one wonders how else the subject could have been taught.

The Education Code of 1871 widened the curriculum. A special grant of 14 shillings per head was awarded to upper standards of schools for scholars who passed in two 'specific' subjects, in addition to the three Rs. These 'specific' subjects included mechanics, natural philosophy (i.e. physics) and chemistry, among others.

In 1872 the Science and Art Department made collections of science apparatus suitable for teaching and these collections were circulated to various schools.

In order to make the teaching of elementary science meaningful the London School Board arranged for each of its schools to be provided with a small box of science apparatus with which to illustrate the lessons. Later a peripatetic demonstrator was appointed. He would travel from school to school, giving lessons each fortnight to pupils in the sixth and higher standards. He would illustrate his lessons by means of demonstrations, using apparatus that he carried from school to school by horse and cart. He would also show teachers how to use the apparatus in the school 'box'. This meant that in between the visits of the science demonstrator, teachers could continue the instruction given. The scheme proved so successful that it was extended to other parts of London and three additional science demonstrators had to be appointed. Usually the demonstrators gave between 18 and 20 demonstrations a week. In 1885 four peripatetic demonstrators were giving science lessons to 2000 children in London schools.

The example set by the London School Board was followed by many other large towns and cities, such as Manchester, Liverpool and Birmingham. The Liverpool School Board instituted courses in elementary physics for boys in Standard IV of its schools. An interesting feature of the Liverpool Board Schools was the total abandonment of textbooks. Instead, guided by an expert demonstrator, pupils were encouraged to take an active part in the science (oral) demonstrations. The peripatetic demonstrator usually travelled to about 20 different schools a week giving four lessons a day, his apparatus being trundled along on a hand barrow. The more advanced students even had the facilities of a central laboratory.

In Birmingham, science was taught systematically by a peripatetic demonstrator and two assistants. They visited each of the schools in turn, giving lessons of two hours and 30 minutes once every fortnight in elementary mechanics and electricity to the boys in Standard V and above. (The girls did home economics and physiology.) Here, too, a central laboratory was built for the use of more advanced students. In one of their schools the Board erected a laboratory complete with lecture room to seat 80 children. About £400 was spent on apparatus to serve schools. A science demonstrator was appointed who prepared his experiments and stored his apparatus in the central laboratory. He then visited various schools, a lad conveying the apparatus on a hand cart. The apparatus would be set up on a trestle table in front of the class, and the demonstrator would perform 'like a conjurer with his box of tricks'.

In between the fortnightly visits of the demonstrator, the regular class teacher, working under the direction of the demonstrator, would give at least one (usually a recapitulation) lesson. In the course of time sufficient apparatus was available to allow some to be kept at each school for a short period, during which time the students could repeat the experiments with the assistance of their class teacher. By 1881 1200 children were being taught science like this in Birmingham schools. By 1889 all boys in Standards V and VI were receiving instruction in either mechanics or magnetism and electricity, whilst girls were taught aspects of physics and chemistry under domestic economy.

With a view to encouraging the establishment of schools which would give not only regular systematic instruction in science, but science instruction of a more advanced nature, the Science and Art Department offered, in 1872, attendance grants to those schools that would follow the conditions set out in the Science and Art Directory. Many elementary schools and many private and grammar schools, tempted by the generous grants, accepted the conditions. So came into being the so-called **Organized Science Schools** which were to play so large a part in establishing science instruction in English education. Such schools were required to provide a three-year course in science, which devoted not less than 13 hours a week to science.

As there was little or no provision for secondary education for the poorer classes, many Standard VI students tended to remain on in extra-standard classes set up in elementary schools. Many progressive School Boards, especially in large towns, added on classes such as a Standard VII, to their elementary schools and so became 'higher grade' schools, higher central schools (such as Sheffield), higher elementary or higher standard schools. The Central School at Manchester developed as a higher grade school out of the former Board School. One room was specially set up for science, it had a good laboratory for 24 students, and a teacher was appointed full-time to teach boys who had passed Standard VI science, which included mechanics, sound, light, heat, magnetism and electricity.

To make room for the higher classes the junior classes were often dropped and pupils only admitted after passing Standard IV. In this way, out of schools established for primary education, a new and progressive form of secondary school came into being. The secondary part of their work could as a rule only be financed by turning the upper classes into an organized science school and so earn grant from the Science and Art Department.

Although both the Science and Art Department and the Committee of the Council in Education emphasized that science teaching must be given a practical bias, few schools possessed laboratories. Indeed, few possessed anything but the scantiest of apparatus, even though £75 would provide all the apparatus needed to teach the popular subject of mechanics. In 1880 the Science and Art Department stipulated, as a necessary condition for grant, that practical work must be done and insisted on provision of proper apparatus. Many school authorities purchased just enough apparatus to meet the requirements, which would then be stored in a cupboard, often gathering the dust of weeks.

The following schedules of lessons for the various standards or classes of the school provide information on *what* was taught as part of elementary science:

Standards I, II, III:	Common objects such as familiar animals, plants and substances employed in ordinary life.
Standard IV:	More advanced knowledge, including the simpler kinds of physical and mechanical appliances, e.g., thermometer, barometer, lever, pulley, wheel and axle, spirit level.
Standard V:	Chemical and physical principles involved in one of the chief industries of England. Physical and mechanical principles involved in the construction of the commoner instruments and of the simpler forms of industrial machinery.
Standards VI and VII:	The preceding, but in greater detail.

In 1890 the Local Taxation (Customs and Excise) Act was passed. The large surplus of customs and excise duties collected was handed over to the newly elected local authorities to spend on technical education. This 'whisky money' as it was called, amounted to some £800 000 a year, and was granted to local authorities to: a) *supply* technical education, i.e., to establish schools and pay the salaries of the teachers; b) *aid the supply* of technical education by making grants to schools and classes giving technical education; c) *promote* technical education by establishing scholarships and exhibitions and by paying the fees of scholars.

The local authorities began to spend lavishly. Technical schools and colleges were established in all large towns. This aid enabled many higher grade and secondary schools to build and equip science laboratories. Between 1877 and 1902 some 1100 school laboratories were built. Since grants were now available from 'whisky money' the Science and Art Department suspended its own grants in 1892.

At the time there was growing concern at the over-emphasis of science in certain schools at the expense of the literary subjects. The curricula of the science schools were as educationally unbalanced as those of the classical grammar schools. Huxley himself argued against this over-emphasis on science. Whilst admitting no education could be called liberal which included only literary studies, he stated quite firmly:

> Nevertheless I am the last person to question the importance of genuine literary education, or to suppose that intellectual culture can be complete without it. An exclusively scientific training will bring about a mental twist as surely as an exclusively literary training. There is a kind of sect, or horde, of scientific goths or vandals, who think it would be proper and desirable to sweep away all other forms of culture and instruction, except those in physical science ... This is not my view.[21]

The Bryce Commission argued that the traditional literary forms of education should not be eclipsed by scientific and technical education. In 1895 literary and commercial subjects were made compulsory in the organized science schools. Part of the grant awarded on the Inspector's report was to be based on the work done by the whole school, both literary and scientific. In this way the one-sidedness of science schools was remedied.

In 1897 payment by results disappeared from elementary schools. In 1898 the Department issued a detailed 'Syllabus of Instruction in Experimental Science' (see Appendix 3) for higher grade (science) schools, which tended to standardize science teaching throughout the country. In the second year of the course not less than eight hours and in the third year not less than ten hours were to be given to science. At least half this time was to be for individual experimental work.

During the last decades of the century science teaching – both content and methods – was being discussed on all sides. The heuristic method of Professor Armstrong was in the air. This is the method (also called the discovery method) of science teaching in which the learner is put in the position of an investigator, the teacher so manipulating the subject matter and activities that the learner discovers facts *for himself*. The British Association for the Advancement of Science, always deeply concerned with the teaching of science in schools, formed a committee to report on the methods of teaching science. The committee emphasized (1898) the importance of practical work by the pupils themselves to such an extent that teachers and school authorities began to feel that teaching by demonstration was of no value whatever. The tradition of teaching science along object lesson lines began to give way to more enlightened methods. The heuristic method of Armstrong came to be adopted in many schools. But this was not easy in elementary schools because of lack of equipment and because of large classes: the only effective method was the discredited one of demonstration by the teacher. The tendency in the later decades of the century was to replace physical science by nature study, which was supposed to lend itself to heuristic teaching without expensive apparatus or specially equipped rooms.

In 1898 elementary science was taught in 2143 'departments' compared with 17 049 for geography, 5780 for history, and 13 456 for English. Even so, as Table 8.1 indicates,[22] there had been considerable growth in the teaching of elementary science in the public elementary schools during the last decade of the century.

Table 8.1

Number of departments in day elementary schools in which elementary science was taught

1890	32	
1891	173	
1892	788	
1893	1073	
1894	1215	
1895	1396	
1896	2237	
1897	2617	
1898	2143	
1899	21301	This marked increase arises from alterations to the Elementary Code in 1895, 1896 and 1898.
1900	19988	

By the end of the century considerable success in furthering the teaching of science was also being achieved in evening schools. More than 10 000 classes were being assisted by the Science and Art Department. There were 78 independent science schools of secondary rank. The very success of the Department's attempts to encourage science teaching had by now made the whole system too large and unwieldy and difficult to operate. Over 190 000 scripts had to be marked at each examination, not to mention the supervision of over 14 000 practical examinations. The Department had played a most significant part in introducing science instruction into the country. It was time now to hand over this responsibility to other bodies that had arisen. By the Board of Education Act of 1899 the Education Department, the Department of Science and Art and the Charity Commissioners were unified under a Board of Education.

Chapter Nine

Physics in the Public Schools during the Second Half of the 19th Century

In 1868, in an address to the South London Working Men's College, of which he was honorary principal, Huxley lambasted the public schools.

> Up till within a few years back, a boy might have passed through any one of the great public schools with the greatest distinction and credit, and might never so much as heard that the earth goes round the sun; ... and as for Science, the only idea the word would suggest to his mind would be dexterity in boxing.[1]

Tyndall, speaking about the neglect of physics in the schools remarked:

> A few days ago, a Master of Arts, the recipient of a modern education, stated to me that until he had reached the age of 20 years he had never been taught anything regarding Light, Heat, Magnetism or Electricity ... I ask you whether this land of old and just renown has not a right to expect from her institutions a culture more in accordance with her present (industrial) needs than that supplied by declension and conjugation.

In 1862 Faraday who, as a youngster, had never been near a public school, was asked to give evidence to the Public School Commissioners. This is what he said.

> That the natural knowledge which had been given to the world in such abundance during the last 50 years, I may say, should remain untouched, and that no sufficient attempt should be made to convey it to the young mind, growing up and obtaining its first view of these things, is to me a matter so strange that I find it difficult to understand.

Professor Armstrong was very critical of what the public and endowed schools – and the universities, the 'classical trade unions' as he termed them – were doing in respect of science teaching.

> In the course of a century or so, the introduction of the experimental method has led to the most extraordinary advance in knowledge.
>
> But, as yet, no organised effort has been made to put our youth in possession of the new knowledge ... forced still to concentrate attention on pious Aeneas and on Caesar, they hear nothing of the great engineers, of Black, Cavendish, Dalton, Darwin, Faraday and many others, who in modern times have made the world of today what it is ...[2]

On another occasion Armstrong commented that 'it would be well if we could destroy the whole wretched academic show and start afresh.'

Science was looked upon with contempt by many of the public schools, even towards the close of the 19th century. In 1869 Hayman, the headmaster of Rugby, was arguing that though science might have a place in the education of the lower orders, it was beneath the dignity of his pupils. Where some (inadequate) provision was made for science the science courses were what Armstrong termed a 'kind of refuge for the destitute', suitable only for those pupils judged inferior in intellect, the more able students encouraged to study the classics and mathematics as being more worthy subjects. At Eton in the 1860s there were 24 Classics masters, eight Mathematics masters, and three toyed with all the other subjects. The scholars proceeding to the universities were quite unprepared in and for science. Only four of the public schools taught some science. A.F. Leach, writing in the latter half of the 19th century, states that at Winchester College 'in my time [about 1863 onwards] Classics were the be-all and the end-all of our education.'[3]

Dr Edward Fox of Balliol recalls that when he got his First Class in the Oxford Science School he wrote confidently to his old headmaster of Shrewsbury to ask for a half-holiday for the boys. The headmaster, Kennedy, replied that he rejoiced in Classical triumphs, was gratified by mathematical honours, pleased with athletic success, congratulated Fox on achieving such success as he had deliberately limited himself to, but to ask for a half-holiday for science was – well, a most unwarranted proposal.[4]

The Public Schools (Clarendon) Commission 1861–4

In response to the general dissatisfaction with what was being taught in the Public Schools, Parliament, in 1861, set up a Commission 'to inquire into the Revenues and Management' of the public schools and 'the studies pursued and instruction given therein.' The members of this Public School Commission were well-chosen, comprising the best of English diplomats and statesmen, learned representatives of the universities and of the public schools themselves.

Although they mustered fewer than 3000 pupils in total, and of these more than 60 per cent were at Eton, the public schools carried immense prestige; and since the governing classes were recruited almost entirely from these schools, their efficiency was a matter of national importance. The Earl of Clarendon warned there was a danger, if adequate steps were not taken, that the new proprietary schools being established would place 'the upper class in a state of inferiority to the middle and lower classes'. The Commission, known by the name of its Chairman as the Clarendon Commission, sat from 1861 to 1864 and carried out a full-dress inspection of the great public schools.

The Commission reported in 1864. As Archer states, the fact that the commissioners, holding as they did that the classics were the best of all educational instruments, condemned the narrowness of such an education, is all the more remarkable. The Commission recognized there were many difficulties in the way of introducing science into the schools. Apart from the heavy bias towards the classics, there was the lack and scarcity of suitable textbooks for school use, generally little encouragement was given to science by the universities, and, above all, there was the difficulty of getting able science teachers. Faraday had to remind the Public Schools Commissioners of this last when he pointed out:

> You want men who can teach science and that class has to be created ... I am often asked what men to recommend to spread a knowledge of natural physical things. I cannot recommend such men; such men hardly exist at present ... If I give a juvenile lecture I lecture in as plain terms as I can to children, and I simplify as far as I can the ideas, but men are not educated to do that as yet. The ordinary schoolmaster does not know how.[5]

H. Cole, Secretary to the Science and Art Department, pointed out that the public schools would only opt for science if the universities made it worth the while of the schools.

> Until increased honours are awarded at Universities for acquiring experimental science, and the knowledge is felt to be remunerative, experimental science will stand no chance with the study of classical learning.[6]

In 1851 the Natural Science Tripos had been established at Cambridge and two years later the Honours School of Natural Science at Oxford. These two events undoubtedly helped to prepare the way for the inclusion of science into the school curriculum. Following on the Royal Commissions looking into the universities of Oxford and Cambridge the two universities endowed scholarships, exhibitions and prizes for public competition. The award of scholarships and exhibitions for science by the universities encouraged the great schools to take up science more seriously.

Rugby

Despite the difficulties of introducing science into the public schools, beginnings were made here and there to introduce some science to them. To Rugby School must go the honour of being the first among the great public schools formally to recognize science as a subject of the curriculum. As the Report of the Public Schools Commission points out:

> Rugby School is the only one amongst those constituting the object of the present inquiry in which Physical Science is a regular part of the curriculum. At Winchester, at Eton, at the Charterhouse, and at Harrow, opportunity has for some time been given to the boys to hear lectures in that branch of knowledge, or to pass voluntary examinations, or to compete for prizes. But in the system at Rugby there are these distinguishing features, that it is assisted by the appointment of a University Graduate residing at the School to give lectures in it constantly; that a particular time is set apart for these, and taken into the usual School hours; that the subject forms a substantial portion of the periodical examinations, the result of which tells in the same degree as Modern Languages on promotion in the School; that the time-honoured distinctions of first and second classes have been conceded to it in the Christmas examination; and that in the contest for exhibitions in the University a paper containing questions upon it is set before the candidates.[7]

Physics was first introduced into Rugby in 1849 with the appointment of a special tutor for the benefit of any boy whose parents might wish him to learn the subject. The lecturer, Dr Sharpe, FRS, was a resident physician in the town. His lectures, which included physics and chemistry, were delivered in the Town Hall and attendance was voluntary. Dr Cilbourn (headmaster) later appointed successively two of the Classics masters who had some taste for natural philosophy.

In 1855 Dr Temple, who had been principal of Kneller Hall, the government teacher-training college, from 1850 to 1855 and who had drawn up official lists of science apparatus for the Department of Science and Art, was appointed headmaster at Rugby. Under Temple science was to be made compulsory for all boys below the upper school. After Temple became headmaster, the science work was undertaken by one of the mathematical masters, and in 1859 the Trustees 'added a lecture room and laboratory to the School building, thus giving considerable impulse to the new studies'.[8] This provision of a Physical Science Lecture Room and Laboratory was a considerable innovation at the time.

In that same year James Wilson was appointed science master. Wilson was a graduate of St John's College, Cambridge, and senior wrangler. However, he had suffered a nervous breakdown and had lost his memory. The most lucrative offer he could get was as a science master, even though he knew no science, apart from hearing Ben Horne's lectures on optics and dynamics – 'as good as lectures could be'.

At that time public school headmasters had to choose between appointing non-scientists (but gentlemen) to teach science or to appoint men of different social standing from their regular staff, who then often had to endure an uncomfortable time in the Common Room. Wilson overcame the handicap of knowing no science by working during the summer in the Birkbeck science laboratories at University College, London, under Williamson and Carey Foster. Wilson agreed that if he failed to teach science satisfactorily he would become a form master (teaching classics) or, failing that, a mathematics teacher. Although he had been appointed as science master, Wilson spent more of his time teaching mathematics. Despite this, his mere presence at Rugby as a science master was a milestone. For a public school to take a science master on its staff was unique. Up to this time what science instruction was given in such schools was usually restricted to *ad hoc* lectures and demonstrations given by peripatetic lecturers, who spent about three weeks in a place lecturing to the local populace, and when possible the lectures would be duplicated at a school.

Rather fewer than 50 boys took up science at Rugby, and their fees, five guineas per annum, formed Wilson's salary. As yet there was no laboratory so, as he states:

> I taught in the cloakroom on the ground floor in the Town Hall, which was furnished with tables and chairs and a cupboard containing apparatus for illustrating elementary optics, hydrostatics, mechanics and heat, and some good electrical apparatus.[9]

Wilson described the position of science in his evidence to the Public School Commission.

> Boys in general are not admitted to lectures in Natural Philosophy until they reach the Middle School. The present Teacher has established this practice in

> the belief that boys before the age at which they commonly reach that point in the School are not well qualified for it. Nor are boys in any part of the School compelled to learn it. It is, in fact, regarded as a substitute for Modern Languages, to which parents may have recourse if they think fit. This alternative, too, is encumbered with the condition that an extra fee of £6 6s per annum, not required for the teaching of Modern Languages, must be paid for instruction in Physical Science. It is formally permissible, however, to learn both Modern Languages and Physical Science, and some boys at the present moment actually take instruction in both, but the practice is discouraged, probably as being supposed to distract the mind with too many pursuits.[10]

Eton

The Report of the Public School Commissioners (1864) states that at Eton 'Physical science is not taught.' However, as at Rugby, and in the same year, 1849, science lectures had begun. These lectures consisted of talks given by men of science invited to the school. This became accepted practice in some other public schools as well. At Eton

> Lectures are delivered once a week during the two winter school-times by men of eminence, on scientific subjects. The attendance is purely voluntary, and there is a small payment (10 pence a week) which defrays the lecturer's charges and incidental expenses. We are told that about one hundred attend. At the end of each lecture some questions are proposed, to which those who are disposed to do so prepare written answers, for the best of which a prize is given; and at the end of the course questions are again proposed, to be answered from recollection. The list of subjects ... appears to have covered a pretty extensive range, and embraced such portions of natural and experimental philosophy as were thought likely to be interesting to boys. Mr. Hawtrey (the Mathematics Master at Eton) judging from his own observation, expresses his opinion that these lectures have been useful in imparting information and awakening interest in scientific subjects.[11]

The eminent men of science who gave the weekly lecture series at Eton in 1849 were Professor Baden Powell and Professor Tyndall.

Harrow

The system of education at Harrow was much the same as at Eton – predominantly classical. However, when Dr Butler came to the school he introduced a weekly mathematical lecture. Butler, who had been a senior wrangler while at university, was admirably qualified to do this. As at Rugby and at Eton, science lectures were given at Harrow. Mr Walker, the well-known peripatetic lecturer in natural and experimental philosophy, occasionally gave these lectures. In addition to the lectures, a textbook was used and voluntary examinations were held in geology, botany, chemistry and electricity, with prizes of five and three guineas. But it appears, 'few entered' for the examinations. In due course, as a result of the interest of several masters, science came to gain more recognition.

Charterhouse

At Charterhouse a voluntary class in physical science was held. In 1848 the school had some successful candidates in the London matriculation examination and since science was a compulsory subject in the examination, the school must have had classes in science.

Winchester College

A.F. Leach states in his *History of Winchester College* that science was introduced into Winchester 'under strong pressure from the Oxford University Commissioners of 1857'. In their letter to the Governors of Winchester College, the Commissioners pointed out that

> to them it appears that good elementary instruction in physical science is most essential in the case of many boys, desirable in all cases, and perfectly compatible with a first-rate classical education. The object might be effected without prejudice to other studies, by setting apart two or three hours every week for lectures in the physical sciences, by putting good elementary works on the subject in the hands of the boys, and by examining them on the lectures once at least in every half year.[12]

In order to help Winchester College to do this, the Commissioners proposed converting the next three vacant fellowships for three men of science who were to reside at Winchester and teach the boys science. This proposal was eventually dropped because, in the view of the headmaster, Dr Moberly, 'except for those who have a taste and intend to pursue the physical sciences as amateurs or professionally, such instruction is worthless as education.'[13]

However, in order to satisfy the Commissioners, it was agreed 'to engage, from time to time, the best lecturers of the day in the various branches of physical science to come to Winchester and give the Scholars successive courses of lectures.' Such lectures were given on Saturday afternoons during the Summer term. The subjects of these courses of ten or twelve lectures were chemistry, geology, electricity, heat and the constituents and properties of water and atmospheric air. All the boys were required to attend, 'but it appears that attendance was not enforced very strictly. There have been no examinations, and no prizes or awards for attention to, or proficiency in this subject.'[14]

Moberly regarded the system of inviting eminent scientists to speak as more satisfactory than specifically employing a science master, since the salary that could be offered a science master would not be sufficient to attract any person of repute, and, in any case, the teacher would tend to get into a rut and stagnate instead of keeping up with the latest scientific developments.

Leach gives an amusing account of these lectures:

> If the French lessons were a farce, the so-called science was even worse ... All Scholars and Exhibitioners were made to attend a science lecture once a week. Nobody else was obliged to attend, and others having to pay extra for doing so, did not attend. Until I got an Exhibition in 1865, I did not even know that there was any such thing as science teaching in the school. Then I had to attend. The lecturer was shared by us with Harrow, and came over once a week. What his

name was I forgot, if ever I knew. But the lecture was invariably on this wise. 'You see these two pith-balls?' No, sir,' said some wag, with his fingers in his eyes, 'I don't see any.' Or, as a variant, somebody would say, 'Two, sir? I see three, sir,' and so on. There was an examination at the end of the term; but it was a pure farce. I obtained eight marks out of 100, and was never a word the worse. The amount of instruction in science derived was probably rather less than under the 'ten lectures a year'.[15]

There is no evidence of any science teaching in the other great public schools Westminster or Shrewsbury, nor at other public schools such as St Paul's, Marlborough and Merchant Taylors'. Dr Kennedy, the headmaster of Shrewsbury School, did say to the Public School Commission that 'There are in the School some models and diagrams in natural philosophy, but no experimental apparatus; and I fear there would hardly be time or staff to work this unless the numbers were much increased.'[16]

The evidence given by such eminent men as Sir Charles Lyell, FRS, the famous geologist, by Michael Faraday, and others, prompted the Public Schools Commissioners to speak out bluntly for reform.

If a youth, after four or five years spent at schools, quits it at 19, unable to construe any easy bit of Latin or Greek ... almost ignorant of geography and of the history of his own country, unacquainted with any modern language but his own, and hardly competent to write English correctly, to do a simple sum, or stumble through an easy proposition in Euclid, a total stranger to the laws which govern the physical world and to its structure ... his intellectual education must certainly be accounted a failure ... yet this is much more commonly than it ought to be the product of English public school education.[17]

The Commissioners were struck by the neglect of science in the great public schools.

Natural science, with slight exceptions, is practically excluded from the education of the higher classes in England. Education with us is, in this respect, narrower than it was three centuries ago, whilst science has prodigiously extended her empire. This exclusion is, in our view, a plain defect and a great practical evil.[18]

Even in those public and endowed schools where science had been introduced, the time allotted to it was wholly inadequate. The Commissioners recommended science as part of a regular course of study for all public schools: 'we are convinced that the introduction of the elements of natural science into the regular course of study is desirable.'[19]

The Commissioners believed that elements of science can and should be taught even to young boys.

The study of Natural Science should, we think, be made compulsory upon all boys for a certain portion of their school life ... The evidence which we have taken on the subject [from Faraday Dr. Carpenter, Sir Charles Lyell, Dr Whewell] convinces us that boys may profitably begin the study of some parts of Natural Science at a very early stage....[20]

The subject is one which, from its material nature and its susceptibility of illustration by outward and sensible proofs, is positively more adapted to the natural comprehension of the young than are the more abstract rules of grammar and arithmetic which they are now called upon to remember and apply. We are of opinion, therefore, that boys even in the lower forms of the School may advantageously be permitted to receive School instruction in the elements of Physical Science.[21]

The Commissioners anticipated the objections that would be raised by the old guard to any such proposal.

It may, perhaps be objected that there is not time for such a course of study as we have described, and that it could not be attempted without injury to classics; that the working hours are already long enough; that not more than a certain quantity of work can be put into a certain number of hours, and that a boy's head will not hold more than a certain quantity of knowledge. Until a few years ago, there was no time for mathematics; at Eton, even now, it is deemed impossible to find time for French. Yet scholarship is none the worse, and general education is much the better, for the introduction of mathematics. We are persuaded that by effective teaching time can be found for these things without encroaching on the hours of play; and that room may be made for them, by taking trouble, in the head of any ordinary boy. We are satisfied that of the time spent at school by nine boys out of ten much is wasted, which it is quite possible to economise. Time is economised by increasing attention; attention is sharpened and kept alive by a judicious change of work. A boy can attend without flagging to what interests him, and what he attends to he can generally retain; but without real attention there can be no progress, and without progress no intellectual discipline worth the name. The great difficulty of a public school, as every master knows, is simple idleness, which is defended by numbers and entrenched behind the system and traditions of the place, and against which, if he be active, he wages a more or less unequal war. We are not without hope that, by the changes which we are about to recommend with respect to the schools collectively and separately, this evil may be considered abated.

There would be, as the Commissioners realized, other objections such as lack of textbooks or lack of able teachers.

Some difficulty will be found at first in obtaining good school-books, and competent teachers; but the demand will create a supply. If Oxford or Cambridge men are wished for, they will not long be wanting when a certain number of Assistant-Masterships in great schools are added to the scanty opportunities of gaining a livelihood now open to students of natural science.

Whilst recommending the inclusion of science into the curriculum of the public schools, the Commissioners remained in no doubt as to the primacy of the classical tradition. The classics were to remain a principal subject, but not, as previously, the *only* subject taken seriously. As in the German Gymnasium, which the Commissioners took as their model, mathematics was given second place.

The importance attached by the Commissioners to classics is reflected in their recommendations to the schools: classics with history and divinity were to have 11 hours a week, arithmetic and mathematics three hours, natural science two hours, French or German two hours and music or drawing two hours. Of the total marks 50 per cent were allocated for the classics; the remaining subjects to share out equally the other 50 per cent.

At the time some public schools were experimenting with introducing, alongside the 'classical' sides, 'modern' sides in which modern languages, mathematics, science, history, geography, would be taught. Cheltenham College, for example, had a 'modern' side in which the boys learnt comparatively little Latin and no Greek, but concentrated on science and modern languages. In 1862 Marlborough College had 62 boys (one seventh of its total) in a 'modern' side.

The Commissioners acknowledged that:

> There are many who have peculiar capabilities for scientific studies, to whom it would be of the greatest advantage to receive a higher amount of scientific instruction than would be desirable for the generality of their schoolfellows, and it may be fairly urged that it would be of advantage for such boys to be allowed to drop some portion of their classical, in order to devote more time to other work.[22]

But they pronounced against 'modern' sides.

The Act that followed the Report of the Public Schools Commission was the **Public Schools Act** of **1868**. Basically, this was an Act to free the public schools from their own shackles, imposed by long out-of-date statutes and biases. It empowered the great public schools to alter their statutes so as to permit the teaching of science. More than that, they were required to 'provide and maintain laboratories and collections of apparatus and specimens'. The purpose of the Act was to reorganize the schools and having done that to leave them ever afterwards with the same freedom as they had before.

Introduction of science

As a result of the changes proposed by the Public Schools Commissioners, science came to be introduced into the public schools in a much more organized and systematic way, and on a much larger scale.

The Rev. T. N. Hutchinson, appointed lecturer at Rugby in 1865, provides this picture of the teaching of mechanics at the school:

> The lectures include experimental investigations into the Mechanical powers, with numerous examples worked by the boys; into the elements of Mechanism, conversion of motion, cams, the steam engine, the equilibrium of roofs, bridges, strength of materials, etc. They are illustrated by a large collection of models, and are very effective and popular lectures.[23]

The main method of teaching was a combination of oral teaching and demonstration. As Hutchinson states: 'the lectures are strictly catechetical, and are fully illustrated by diagrams, specimens, models and experiments, as the case may be.'[24] The combination of catechetical lecture and lesson did not include anything like mere dictation of notes.

> The lecture is given, interspersed with questions, illustrations, and experiments, and the boys take rough notes, which are recast into an intelligible and presentable form in note books. These are sent up about once a fortnight, looked over, corrected, and returned; and they form at once the test of how far the matter has been understood, the test of the industry, care, and attention of the boy, and an excellent subject for their English composition.
>
> Examination papers are given to the sets every three or four weeks, and to these and to the note books marks are assigned which have weight in the promotion from Form to Form. The marks assigned to each subject are proportional to the number of hours spent in School on that subject.
>
> There are School prizes given annually for proficiency in each of the branches of Natural Science.

H.F. Newall, who was later Professor of Astrophysics at Cambridge, was given time off from classics in order to attend Hutchinson's lectures and laboratory classes. After six months' experience of introducing science into the Lower School at Rugby, the subject was dropped from the Lower School at the end of 1865, 'the boys appearing hardly equal to the work' in the words of Hutchinson. The reason for the boys being unable to cope with mechanics was basically because of their poor grounding in mathematics, as J.M. Wilson, the science master at Rugby, pointed out in a paper to the British Association for the Advancement of Science on 'Natural Science in Public Schools' in September 1866.

> Mechanics are found rarely to be done well by those who are not also the best Mathematicians. But it is a subject which in its applications interests many boys, and would be much better done, and would be correspondingly more profitable, if the standard of Geometry and Arithmetic were higher than it is. The ignorance of Arithmetic which is exhibited by most of the new boys of 14 or 15 would be very surprising, if it had not long since ceased to surprise the only persons who are acquainted with it; and it forms the main hindrance to teaching Mechanics. Still, under the circumstances, the results are fairly satisfactory.[25]

In this paper Wilson gave interesting advice to schoolmasters contemplating introducing science into their schools. He stated that the decided opinion of those who had given thought to the matter was that experimental physics ought to form the staple of science teaching in schools; he would exclude 'the frightfully crammable' chemistry and geology. Regarding physics he favoured mechanics and heat and suggested two possible methods of teaching them. One is 'to put a textbook into the hands of everyone, require certain parts of it to be learnt, and illustrate them in an experimental lecture with explanations.' This method helps the boy over the difficulties. But:

> what cost the world centuries of thought is told the boy in a minute, and the one power which the study of physical science ought to create, in a word, exercising all the active faculties which are required for an investigation in any matter, these may lie dormant in a class while the most learned lecturer experiments with facility and explains with clearness.

In preference to this he would use the other method, 'the method of investigation' which bears a striking resemblance to what came to be later known as 'heurism'. Wilson concluded that a union of the two methods was probably best. Wilson put his ideas into practice at Rugby, which became a pattern for other schools. His influence was wide and deep.

The Public Schools (Clarendon) Commission had highlighted the 'barrenness of the land' in relation to the public schools. A few years later a Royal Commission on Scientific Instruction and the Advancement of Science, more familiarly known as the Devonshire Commission (1871–5), carried out a most thorough and searching inquiry into the teaching of science in all schools and educational institutions throughout the country, from elementary schools to universities.

Whilst the Clarendon Commission had found that with few exceptions there was little science teaching in the public schools investigated by it, ten years later the Devonshire Commission reported some progress in this direction. The Sixth Report of the Devonshire Commission deals specifically with science teaching in the public and endowed schools.

Rugby School

In the Middle School at Rugby (totalling 270) the boys (average age 16 years 6 months) studied physics, the textbook used being Dr Balfour Stewart's *Textbook on Physics*. This course of elementary physics displaced the former course in mechanics. Because of certain school arrangements some boys of the Middle School also studied astronomy for two terms, using Norman Lockyer's *Astronomy*. At the end of the first year of the boy's work in the Middle School he was allowed to choose between geology and physics for the second year's study. But he could not take physics unless he performed fairly well in mathematics.

The headmaster, Mr Hayman, points out:

> In the middle school, and therefore with younger boys, the teaching of Physics has been found to be less satisfactory than in the upper school. The subjects that have been tried are Hydrostatics and Pneumatics, with the elements of Heat, but, although most of the boys have shown considerable interest in the facts and experimental illustrations, the power of grasping the principles and applying their knowledge to easy problems has been acquired comparatively by few.[26]

In the Sixth Form and Upper School at Rugby, totalling 180 boys of average age around 17 years, chemistry and physics were the only science subjects taught, with the exception of a short course of physiology. Under physics was taught 'Heat, Magnetism and Electricity and Geometrical Optics treated experimentally'. Hydrostatics and pneumatics were also occasionally included. The school reported that experience had shown that it was possible to teach physics more successfully to a larger number of boys than chemistry for which 'it would seem that a boy requires an almost special turn of mind to grapple successfully with the ideas involved in chemical affinities and reactions.' Each science subject was given two lessons a week, of about one hour's duration each.

For electricity and magnetism the textbook, 'regarded rather as affording the data upon which a lesson is to be founded', was Ganot's *Physics* (see Appendix 4). For heat, in addition to Ganot's *Physics*, Deschanel's *Natural Philosophy* was used. 'Here, again, abundant experimental illustration is combined with written answers to questions and the working of numerical examples.' For optics, the textbook was Airy's *Geometrical Optics*.

Examinations in science were held from time to time and the marks were used in determining the place of each boy at the end of the term. Moreover,

> In the higher part of the school, a boy who shows ability may be allowed to drop verses, together with a certain amount of general classical work, and devote the extra time to mathematics or Natural Science. The important fact is, that by so doing he does not lower his position in the school or lessen the chance of future promotion, as he receives a full equivalent of marks for actual work.[27]

Wilson favoured the science work at the school (at least in the Upper School) being examined and moderated by an outside university examiner.

In 1859 Rugby established a small laboratory and natural science lecture room. The expense of erecting and fitting the natural science Schools at Rugby was £2000, exclusive of apparatus. By now the staff teaching the ten 'sets' of science classes in the Middle and Upper Schools had been increased to five – the senior natural science master plus four mathematical masters who each gave a portion of their time to science. A competent laboratory assistant had also been engaged, 'part of whose duties consists in superintending the boys, at certain hours, when working at Practical Chemistry'.

What effect did the introduction of science teaching have on Rugby? This is what Canon Wilson had to say in a Paper to the British Association in 1866.

> Lastly, what are the general results of the introduction of scientific teaching in the opinion of the body of Masters? In brief it is this, that the School, as a whole, is the better for it, and that the scholarship is not worse. The number of boys whose industry and attention is not caught by any School study is markedly less; there is more respect for work and for abilities in the different fields now open to a boy; and though pursued often with great vigour, and sometimes with great success, by boys distinguished in Classics, it is not found to interfere with their proficiency in Classics, nor are there any symptoms of over-work in the School. This is the testimony of Classical Masters no means specially favourable to Science, who are in a position which enables them to judge. To many who would have left Rugby with but little knowledge, and little love of knowledge, to show as the results of their two or three years in our Middle School, the introduction of Science into our course has been the greatest possible gain; and others who have left from the Upper part of the school, without hope of distinguishing themselves in Classics or Mathematics, have adopted Science as their study at the universities. it is believed that no master in Rugby School would wish to give up Natural Science and recur to the old curriculum.[28]

Eton

At Eton in 1868 all boys in Form V (B and C Divisions), numbering about 350 in all, and aged 14–15 at the start of the form, were obliged to study some branch of science for two lessons a week. They began with physical geography, and then moved on to mechanics, hydrostatics, heat and elementary astronomy. Mr Madan, who taught science at Eton, pointed out that heat, treated in a very simple way, was very attractive to the younger boys. The textbooks used were Balfour Stewart's *Elementary Physics* and Lockyer's *Astronomy*. Some idea of the standard of work done can be ascertained from the specimen examination questions shown in Appendix 5. For the senior boys the textbooks used were Ganot's *Physics* and Tyndall on *Heat*.

As at Rugby, Eton had a chemistry laboratory (opened in October 1870) but no physics laboratory. However, some 'schools' for physical science were erected near the chemistry laboratory. In 1879 a 'school of mechanics' was built, and a workshop seven years later. In his evidence to the Royal Commission on Scientific Instruction Madan stated that 'practical work in the laboratory is done entirely out of school, the time devoted to it being from 3 to 3½ hours each week (double the time spent in regular school lessons)'. His views as to the relative merits of theoretical and practical science teaching are interesting. In reply to the question 'which form of lesson is found to be most effective?' Madan stated:

> It is not quite easy to compare the oral instruction with the practical work. The latter is strictly supplementary to the former, and serves to fix in the mind what has been previously shown.
>
> It is doubtful whether practical work by itself is sufficient. Without previous oral teaching a boy is apt to do an experiment mechanically, without realising its full meaning. During laboratory work the boys are advised to consult their text-books and note books, and constant reference is made to what has been shown in lecture.

Whereas the tendency in general today would be to base the theoretical (oral) work on what had previously been 'discovered' or ascertained by experiments, for Madan – and presumably for many science teachers of the time – the theoretical instruction formed the main edifice of the teaching, the practical work being regarded as 'supplementary'. Indeed, there were worthy men who thought demonstrations by the teacher quite unnecessary and practical work by the boys a waste of time. Todhunter, when he heard that a young master was setting up the 'guinea and feather experiment' as a demonstration remarked that demonstrations had a demoralizing influence on the boys. 'They should,' he said, 'accept the master's word.' One must remember that Madan was primarily concerned with teaching chemistry and in teaching chemistry much can be said for Madan's point of view.

Like Rugby School, Eton could boast a large observatory. It contained a 6-inch equatorial telescope (donated to the school by the liberality of some of the masters) complete with divided circles, driving clock, etc., a sidereal clock, together with a few necessary books and diagrams. Its total cost was about £600.

> It is under the management of a committee of masters who take it in turns to be in the observatory from 9.30 to 10.30 p.m.

> The names of boys who are recommended by their masters are entered in a book, and four or five are taken up on every fine night, as well as occasionally during the day. [29]

Harrow

The reply given in July 1870 to the questionnaire sent out by the Commissioners for Scientific Instruction informs us that 'experimental physics is the only subject which has been taught in the School.' This physics was taught to all boys (about 40) on the Modern side, in three lessons a week. All boys in the V forms and Upper Remove on the Classical side (about 140) were given one lesson a week. The textbooks used were Deschanel's *Physics*, Maxwell on *Heat*, Tyndall on *Sound* and Lockyer's *Astronomy*. The science master reported that:

> As far as it has been attempted, a lesson in which the boys are made to repeat experiments which they have seen performed by the teacher at a previous lesson, or other simple experiments, is found to be by far the most effective.

Considerable encouragement to science teaching at Harrow was provided by the erection of the chemistry and physics laboratories and science classrooms.

Winchester College

Mr G. Griffith, who was Deputy Professor of Experimental Philosophy at Oxford, reported to the Schools Inquiry Commission (1865) that since 1863 he would go to Winchester once a week to teach science.

> I go there once a week. I have two classes of boys, each class consisting of about 30. I give them lectures upon different branches of physics, and astronomy. By physics I mean mechanics, acoustics, light, heat, electricity, and magnetism. I expect the boys to take notes of these lectures, which are afterwards written out. I inspect the notes and correct them.[30]

When asked if science should be taught at school, Griffith replied:

> Even very young boys may be taught the elements of physics. They can be shown simple experiments, and if they are made to understand the explanations which are given of the experiments and afterwards to write an account of them, I think that they have an opportunity of exercising their mental powers in a way which young boys at school do not often have.[31]

He explained that one of the difficulties of trying to introduce science into the public schools was that 'boys know it pays better at the Universities to give their principal attention to classics.'

By 1869, under Dr George Ridding as headmaster, science at Winchester had changed for the better. In 1870 several buildings were converted to make excellent physics and chemistry laboratories. In the sixth form, those intending a career in the army substituted German for Greek, and 'cultivated science and mathematics at the expense of Plato and Aeschylus'.[32]

In his reply to the Commissioners' Questionnaire in October 1872, Mr Ridding remarked that, in addition to mechanics and hydrostatics, which were taught to some of the senior boys (aged 17), other branches of physics were also taught. 'The Experimental Courses in Physics interest a good many boys, and give them new ideas. Electricity seems to be the best subject; Chemistry, as taught in courses of lectures, does not seem to answer.'

The apparatus the College possessed consisted of 'mechanical models, elementary electrical apparatus, apparatus for illustrating physics'. The textbooks used were Lockyer's *Astronomy*, Tyndall on *Heat*, Ganot's *Physics* and Balfour Stewart's *Physics*.

Later in his reply to the Commissioners' questionnaire, Mr Ridding listed these factors as the principal obstacles to teaching science at Winchester.

> Those incident to a period of transition:
> 1. The costliness of starting it, especially in such a branch as chemistry.
> 2. The difficulty of finding teachers possessing actual knowledge of practical and experimental kinds, who are also able to manage a class of boys.
> 3. The past feeling of inferiority to other subjects, a feeling, perhaps, hardly past.
> 4. The feeling that it does not 'pay' as well as other subjects.[33]

In response to the question as to how the universities could best assist science-teaching in schools Ridding replied: 'By making it *essential* for *their best men* to know some rational amount of Natural Science.'

Other schools

From evidence given to the Devonshire Commissioners, it seems that physics was taught at **St Peter's College, Westminster (Westminster School)**.

> At St Peter's College, Westminster, applied physics is the only branch (of science) which has as yet been attempted; a course of lectures on Heat seemed to answer best for the classes generally, but the subject of Electricity was more interesting to a few of the best pupils.

It was the view of the headmaster, Charles B. Scott, that instead of trying to cover all subjects in mediocre fashion it was better to concentrate rather on a few: to do this schools should be divided in order to specialize either on science or languages.

> I can only repeat my conviction that if science teaching is to be really efficient as an instrument of education, it must be made the principal subject of study; and that schools should be divided into classes, some taking science, and others language, as the basis of their system ... adding in each case such further studies as time will allow to be pursued with advantage. The notion of pressing all subjects on all schools is in my judgment fatal to any efficient training.[34]

At **Charterhouse School** by 1870 science was an obligatory school course for all boys from Form III up to the Sixth Form.

Chapter Ten

The 'Lesser' Public Schools

Besides the seven 'great' public schools mentioned in Chapter 9, there were well over 700 grammar schools of every grade and class and merit, all over England. Many of these were of 'public school' rank, for example, University College School, London, the City of London School, King's College School, London, Marlborough College, Mill Hill, Stonyhurst, Rossall, Clifton College, Cheltenham, Uppingham, Taunton, Dulwich, Wellington, and the grammar schools in Manchester and Bradford. Many of these minor public schools were not shackled by outworn statutes that restricted their curricula to the classics, and so gave instruction in natural and experimental philosophy, i.e. physics. Their lead and example was to have great influence on science teaching in the secondary schools of Britain.

Rugby School, which built a small chemistry laboratory in 1859–60 is often cited as the pioneer of science education in English schools. This is not so. Even before Rugby School, classes in experimental physics were conducted at University College School in 1835 and the City of London School had set aside part of a basement passage 'specially adapted for practical science' in 1847. Stonyhurst built and was using a physics laboratory as early as 1808. Even earlier, an old Stonyhurst boy, Richard Shiell, MP, records how a demonstration lesson on electricity he was attending was brought suddenly to an end by the news of Nelson's victory at Trafalgar in 1805.

Stonyhurst College

Stonyhurst College was founded by the Jesuits in 1593 at St Omers in Northern France to educate the sons of English Roman Catholics. It finally established itself in England in 1794. By the middle of the 18th century the physical sciences were taught, as far as they were known, in Jesuit schools and 'physical cabinets' (collections of apparatus) were in regular use. The *Ratio Studiorum*, the Code of Studies of Jesuit schools and colleges, laid particular emphasis on the study of mathematics and natural philosophy.

Stonyhurst College consisted educationally of:

1. A school of about 200–250 boys. Those boys at the top of the school (in Rhetoric, aged 16–17) took the Matriculation Examination of the University of London.
2. A college for the general education of young men – young 'lay philosophers' – averaging about 25 in number, who took courses leading to degrees of the University of London and also Civil and Military examinations such as for the Indian Civil Service (established 1855) and the entrance examinations for Sandhurst and Woolwich (established 1857).

Stonyhurst College was using a physics laboratory as early as 1808, though whether this was for use only by the young gentlemen or by senior boys is not stated clearly. For teaching purposes Father Simpson purchased a considerable store of apparatus, to assist his 'teaching of mechanics, of the Newtonian system of gravitation'. The laboratory was also furnished with 'an apparatus of capital instruments for lectures on astronomy'. A visitor to the college in 1824 wrote

> The Philosophical room occupies the eastern wing of the new building. It is appropriated to the Philosophical apparatus ... a handsome room forty-eight feet long, thirty-three feet broad by twenty high. ... it contains two large, handsome glass cases, one of which is stocked with instruments illustrative of the Mechanical, Hydrostatical, Hydraulic and Pneumatic principles among which is a most beautiful and complete model of a Steam Engine, the other is fitted up with those appertaining to Astronomical and Optical Purposes. Besides these there is a fine Newtonian Telescope, 7 feet focal length and 8 inches diameter, a beautiful highly polished Transit Instrument 4 feet focal length both by Cary, London, a good Astronomical circle, Telescope, Time-piece, Electrical Machines, Air-pump, Large and Handsome Globes, etc., etc. This splendid room opens by two large sliding doors into another room of nearly the same dimensions with rows of seats rising in front up to the ceiling, thus rendering it most convenient for the purpose to which it is allotted, viz. for the Philosophical and Classical Exhibitions, etc.[1]

In 1811 the newly-invented gas lighting was installed at Stonyhurst, the first public building lit by gas. This enabled the chemical laboratory to be put to effective use when no other school in England had one. In August 1812 a course of lectures, with experimental illustrations on heat, was advertised by the master in charge of chemistry.

In 1840 Stonyhurst College became affiliated to the University of London, and so, whilst the great public schools of England were still following a predominantly classical curriculum boys from Stonyhurst were being entered for the matriculation examination of the University of London, with its compulsory papers in science. During the 1840s Stonyhurst averaged about seven matriculation successes a year.

In addition to entering candidates for the London degrees, Stonyhurst also entered pupils for the army entrance examinations to Sandhurst and Woolwich. These examinations contained science papers – experimental science (chemistry, heat and electricity) – though they were not compulsory. Roman Catholics were barred from Oxford and Cambridge by religious tests. These tests were not modified till 1854 and not finally removed till 1871. The matriculation examination and the degrees of the university of London which were open to external students after 1858, cleared the way to many who, for religious or other reasons, could not otherwise have obtained a higher education.

In 1896 the English hierarchy agreed to allow Catholic students to attend the Universities of Oxford and Cambridge, and Stonyhurst (along with other Catholic colleges) broke from the London University examinations (e.g. matriculation) and in 1897 adopted the Higher Certificate of the Oxford and Cambridge Joint Board.

Facilities for science improved in 1902 with the conversion of a room into a physics laboratory for the use of boys in the Grammar class and above. Favourable comment on the science facilities at Stonyhurst at the beginning of the 20th century came from some members of the British Association for the Advancement of Science who came to visit Stonyhurst from their Annual Meeting at Southport in 1903. The leader of the group, Professor Dewar, FRS, said that Stonyhurst was far ahead of public schools in the matter of science and could well serve as a model.[2]

University College School, London

The teaching of applied mathematics, which comprised several branches of physics, e.g. mechanics, hydrostatics, optics and sound, was started at UCS in 1835–6, under a Mr W. Wright BA (later Headmaster of Colchester Grammar School). In 1839 Wright was succeeded by the Rev. William Cook, who taught till 1860. In Cook's second year natural philosophy was no longer a mere alternative to Greek and 'became an integral, though not a compulsory, part of the curriculum'.

The science classes at UCS were remodelled in 1863–4 under Mr C. Haughton Gill, FCS. In his evidence to the Royal Commission (December 1870) he provides an excellent picture both of the content and methodology of the physics taught at University College School.

> During the five years in which I had charge of the classes of experimental science, 1863–4 to 1867–8 inclusive, instruction was given in the following subjects:
>
> 1. **Elementary Chemistry**, taught by means of lectures with full experimental demonstrations; monthly examinations, in which many numerical questions always found place; and daily written exercises on the preceding lecture.
> 2. **Experimental Physics**, taught in the same way as chemistry, and with careful exclusion of mathematical formulae. The subjects taken were:
>   Elementary dynamics and statics, including hydrostatics and pneumatics; so treated as to lead the boys to construct the formulae relating to the simpler cases for themselves from the results of experiments exhibited or described.
>   Elements of heat, including the method of determining the constants and calculations relating to specific heat, etc.
>   Elements of light and radiant heat, including nature of light, simple laws governing its transmission, formation of images in plane and curved mirrors, refraction of light, index of refraction, formation of images by lenses, compound nature of light, nature of colour, etc.
>   Elements of sound, and, when time could be found, some portions of electricity and magnetism.

When Gill resigned in 1868 the senior class in natural philosophy was taken by Mr Walker, who taught applied mathematics. From the communication to the Royal Commission on Scientific Instruction by Mr J. Walker we get an excellent picture of the subject matter and methodology of physics he taught at University College School. The references to his 'experimental' teaching of the subject are noteworthy.

> In the current session, 1870–1, two lectures in each week are given of an exclusively experimental character, for the purpose of establishing fundamental physical laws, and the properties of the simple machines and instruments; while two other 'exercise lessons' are given in an ordinary schoolroom, away from apparatus and models, in which the boys are practised in easy deductions from the fundamental principles established in the experimental lectures, and in the solution of such questions as usually form a part of an examination in the elements of natural philosophy ... Generally, in science teaching, I make it a practice not to state anything which can be drawn from the boys themselves by questions more or less 'leading' according to the nature of the deduction sought, or by appeal to their own observation and experience.[3]

Especially noteworthy is Walker's use of a methodology of science teaching that would be adjudged excellent today, *deducing* the laws and facts of nature *from* the pupils (rather than telling them) as a result of observations arising from experimental work.

Also worthy of note is his use of 'practical examinations'. At the examination of the natural philosophy class towards the end of the session, selected boys from the class are required to repeat certain experiments and demonstrations with the apparatus with their own hands.

From the session 1868–9 chemistry and chemical physics were taught by Mr T.A. Orme, FCS, author of a celebrated *Science of Heat*. In his evidence to the Royal Commission on Scientific Instruction, Mr Orme stated: 'The object sought, whilst giving instruction in these subjects, has been rather to teach boys to study scientifically than to impart to them any purely technical information, which would only in a few rare cases be of service to them in after life.'[4]

Orme's emphasis on his pupils acquiring a scientific attitude and habit of mind rather than logging up an encyclopaedic store of soon-to-become-obsolescent facts would meet with any pedagogue's approval today.

King's College School, London

Physics was being taught at King's College School in 1844. In the same year Professor Daniell, of King's College, was giving a course of lectures in chemistry. In 1845 Charles Tomlinson, a prolific writer of physics textbooks, was the lecturer in experimental science. This science instruction was given partly in the school and partly in the lecture rooms of the College.

In his evidence in February 1866 to the Schools Inquiry Commission, G.W. Dasent, Examiner for the Council of Military Education and for the Indian Civil Service, pointed out in forceful terms the uselessness of teaching science merely by lectures: 'It has generally resulted in teaching nothing at all. It is a very pretty way of spending time.' As an examiner Dasant attended the chemistry lectures that Professor Daniell, of King's College, gave to the boys of King's College School.

> They were excessively delighted; but when they came to be examined they knew nothing at all, though Mr. Daniell did everything in his power. I attended the class myself, so I know the bad effects of it. That is the fault of lectures in general. It is very easy to delude yourself at a lecture that you are learning a great deal, when the fact is you are learning nothing.

Cheltenham College

At Cheltenham the College was divided into a Classical side and a Military side. At first science was not taught in the Classical Department, where the boys were prepared specifically for scholarships at the universities and for the Indian Civil Service. However, instruction in 'the chief facts and principles of physical science' was begun in the Military Department in 1854, the subjects taught being geology, heat, chemistry and electricity. Several schools met the needs of boys aspiring to Woolwich and Sandhurst by the creation of 'modern' sides in which mathematics and science were emphasized. These military and scientific departments, at Cheltenham College, Wellington, the City of London School and others, were exceptional instances of schools in which instruction in science held the first place.

The headmaster of the Military Department, 'a man of great experience and great wisdom in teaching, who himself taught science for many years before any special master for science was appointed', outlines the manner of the science teaching.

> The upper classes receive instruction, by means of lectures, three times in the week, and the lower classes twice a week. Besides the lectures, the boys have to read up the subjects from textbooks, and those who are studying chemistry and electricity also work in the laboratory under the supervision of the science master to the military and civil department.[5]

Wellington College

From the opening of the College in 1859 instruction in physical chemistry was given, in a small laboratory, to a voluntary class of boys in the upper school. This instruction in physical chemistry would have included some teaching on electricity and heat, which then formed part of the chemistry curriculum. From 1867 chemistry, heat and electricity came to be taught on a more formal basis to a larger number of boys.

The College reported that it had 'a fairly complete set of electrical apparatus for chemistry, and heat little more than what is absolutely necessary.'

Taunton School

Taunton School is significant in that we find physics – and 'experimental' physics at that – being taught throughout the school. This was undoubtedly due to the influence of the Rev. W. Tuckwell, the Headmaster, who, like Canon Wilson of Rugby, was a great champion of science teaching in the schools.

In August 1869, Tuckwell offered the fruits of his experience in teaching science in a paper he read before the British Association at Exeter. The Devonshire Commissioners appended this paper on *The Method of Teaching Physical Science in Schools* to their *Sixth Report*. It is quoted here fairly extensively not only for its wisdom but because of the great influence it was to have on other schoolmasters who were introducing science into their schools.

> The claims of Physical Science, on *a priori* grounds, to a fair place in the course of school work, have been abundantly vindicated, and are, I suppose, established. But the method and details of its teaching, the books and apparatus which it requires, and the amount of time which must be given to it, are points which can be decided only by experiment, and have not yet been decided at all. I cannot premise too distinctly that the aim of this paper is practical. Of the necessity for teaching science to their boys many good schoolmasters are convinced; as regards the machinery by which it is to be taught, they mostly confess their ignorance, and cry aloud for guidance. In my own school it has been taught systematically for the last five years, and I offer the fruit of this experience, very humbly, to all who are interested in education.
>
> The subjects to be taught, the time to be spent upon them, the books and apparatus necessary, and the mode of obtaining teachers, are the points on which information seems to be required. I will take them in order.
>
> The subjects which naturally suggest themselves as most essential are Experimental Mechanics, Chemistry, and Physiology.
>
> The laws of light and heat will be taught as prefatory to chemistry. Electricity attracts boys so readily that with very little help they will make great progress in it by themselves.
>
> The time to be given to science should not be less than three hours a week. At this rate two years may be given to mechanics, two years to chemistry, one year to botany; while the rest, if any remain, will be free for physiology.
>
> For the first year the teaching may be *viva voce*, with easy problems and abundant experiment; care being taken that each week's lectures shall be reproduced on paper, and great attention being paid to correct drawing. In the second year the teaching will be more minute and more extended, and a good book will be mastered. At the end of this time the class is fit to pass creditably the Oxford Local Examination for juniors, and has done with mechanics for the present. The third and fourth years will be given to inorganic chemistry.

The subject of books and apparatus, involving as it does the question of expense, is of the highest practical importance. Apparatus need not cost much.

We have taught mechanics efficiently, that is to say, we have passed our classes for the last three years in the Oxford Local, with a good air pump, a set of pulleys, models of the force pump, and the common pump, and with the diligent use, during the second year, of Newth's *Natural Philosophy*. But we have lost no opportunity of making the boys acquainted with machinery; from the crane and the water-mill of our daily walks, to the steam engine and the spinning jenny of the manufactory; for he who has not examined engines at work will never understand them clearly, or describe them correctly.

It remains only to examine the mode of obtaining teaching power. Institutions which can give so high a salary as to command a London bachelor of science or a first-class Oxford or Cambridge man, will find no more difficulty than attends the choice of all masters: where this is not the case, it is sometimes possible, by combining mathematics with physical science, to tempt a superior man with a sufficient income; and if only a small salary can be given, the ordinary pass B.A. of the London University will sometimes make a fairly good teacher. But one point has struck me forcibly in my own experience; namely, the unexpected value of general culture in teaching special subjects. The man who knows science admirably, but knows nothing else, prepares boys well for an examination; but his teaching does not stick. The man of wide culture and refinement brings few pupils up to a given mark within a given time, but what he has taught remains with them; they never forget or fall back. I am not sure that I understand the phenomenon, but I have noted it repeatedly.

Secondly, the teaching of science makes school work pleasant. The boys' evident enjoyment of the scientific lesson rouses the emulation of other masters. They discover that the teaching of languages may become as interesting as the teaching of science. They realize – a point not often realized – the maxim of Socrates that no real instruction can be bestowed on learners by a teacher who does not give them pleasure.

Amongst the papers Tuckwell gave to the Devonshire Commissioners was one dated 1 May 1865, addressed to Tuckwell from Professor John Phillips, of Oxford University. In it Phillips pointed out how scientific principles could be taught to young people through practical means rather than overtaxing immature minds with grand theory.

How much fitter for a boy is the guinea and feather experiment than the 'theory of gravitation'; one peep through a telescope, than the 'laws of dioptrics and catoptrics'; the burning of iron in oxygen, than the 'atomic theory'! These laws and theories they may fully master in after years, and all the more easily, that they have seen some of the characteristic phenomena while the eyes are bright and the doors of memory are open to new impressions.

Clifton College

The Rev. J. Percival, the headmaster, provides a good idea of the teaching of science at Clifton in the year 1870. In his Memorandum to the Devonshire Commissioners, dated 19 October 1870, he writes:

> Before the year 1867, Natural Science was taught only as an extra subject, and by lectures at which attendance was voluntary. But since the spring of that year, scientific instruction has formed part of the regular routine throughout the whole College, with the exception of the two lowest forms of our Junior School, which consists of boys from 9 to 11 years of age [to whom no science was taught] and the two highest forms on the classical side [Forms V and VI], in which science is an optional subject.

In his Memorandum Percival reported that the introduction of science to Clifton had resulted in:

1. A new interest in their school work to many boys who before had scarcely any interest at all in it.
2. A consequent increase of the spirit of work throughout the school.
3. The general stimulus which is given to boys of a scientific turn by the opening thus afforded them.

In a letter to *Nature* in 1881 Professor Parker pointed out how doing practical work with a large class of boys could lead to disorder. 'The consequences of setting large classes of young boys to make oxygen, or to take a specific gravity ... each for himself, might prove rather subversive of order than conducive to improvement.'[6] In response, H.B. Jupp, a science master at Clifton replied:

> It may be interesting to some of your readers to know that at Clifton College we have lately tried the experiment of turning some of our ordinary physics classes, numbering from 24 to 30 boys, bodily into the physical laboratory, where they work at weighing, measuring, finding specific gravities and such matters, under the control of a single master. The boys work in pairs, each with a little manual of instruction, and each pair with a separate cupboard of cheap apparatus. Two such classes are taken by Mr. Worthington and myself, and we are both agreed that whatever difficulties we may feel, we have none in the matter of discipline. On the contrary, the boys are with scarcely an exception most keen and eager at this work.[7]

The Mr Worthington referred to above did pioneering work in establishing a suitable syllabus for practical work in physics, based largely on his experience at a school at Saltaire where he had previously been engaged. In 1881 appeared A.M. Worthington's *An Elementary Course of Practical Physics.* Five years later appeared his *A First Course of Physical Laboratory Practice*, reprinted in 1903.

In 1869 Mr J. Angell, from the Mechanics' Institution, was appointed to teach physics. He gave the Commission a very full account of his teaching.

The whole of the time occupied in the class is given (with very few exceptions) to oral testing and experimental demonstration.

I find the 'Socratic' form of lesson the most efficient. In this form of lesson the information required to be taught is either deduced or built up by series of logical questions systematically put to the boys, not on what they have acquired verbally from the book or the teachers, but put to them on the apparatus, the experiment, or the phenomena, or problems as they are actually before them. In this way the observing and reasoning processes, and those of scientific literary exposition are cultivated simultaneously.

A very few only of the upper boys in the Sixth Science form do any systematic practical work, because of the want of a physical laboratory (a great defect). The boys are, however, often called down during an ordinary lesson and required (before the class) to arrange apparatus for a particular demonstration, or to perform an experiment. Boys are often required to construct at home, in the place of their ordinary home lessons, special pieces of apparatus, as – a polariscope, selenite figures for the same, an electrophorus, induction coil, diffraction apparatus, models of Nicol's prisms, electrotype apparatus, voltaic batteries, etc.[8]

Chemistry laboratories were set up by Marshall Watts in 1870–1. Mr Dill, the High Master, recognized the need for corresponding laboratory facilities for physics. The planning and furnishing of the proposed new physics laboratory was entrusted to W. Haldane Gee, BSc, an old boy of the school, and at that time demonstrator and assistant lecturer in physics at Owens College, Manchester.

Dulwich College

According to the evidence of the Master of Dulwich College, Mr Alfred Carver, given to the Royal Commissioners on Scientific Instruction on 7 April 1873, instruction in physics was given at the college. In 1871 62 boys (average age 14 years 6 months) were studying electricity and magnetism (1½ hours a week). 'Instruction is also given from time to time in other branches of science, e.g. heat...' In addition there were courses in the evening (one each week) on light and heat.

Dulwich College was one of the public schools with the vision and foresight to invest in science laboratories. It had two laboratories – the Lower (or Chemistry) Laboratory and an Upper or Small Laboratory for physics and chemistry.

St Paul's School and Merchant Taylors' School, London

As late as 1871 both these schools responded negatively to the Commissioners with regard to scientific instruction at their establishments. The High Master at St Paul's explains that 'We have, in fact, no time, nor indeed space, to extend our system at present, the boys not residing here, and the school accommodation being too much stinted to allow additional classrooms for practical instruction.' Similarly, the Headmaster of Merchant Taylors, gave as the reason that there was 'no provision, at present for what is termed scientific instruction ... It is impossible, from want of room, and from other circumstances, to introduce new subjects so long as the school remains in its present confined locality.'

Chapter Eleven

Physics in the Grammar Schools in the Second Half of the 19th Century

The Schools Inquiry Commission

The Endowed Schools

Whilst the Public Schools Commission of 1864–8 reported on the state of education in the seven leading public schools, and the Duke of Newcastle's Commission in the 1850s on Popular Education covered the education of boys and girls of the labouring classes, it was left to the Schools Inquiry (Taunton) Commission to investigate every other aspect of secondary education provided – by the endowed schools, by proprietary companies and by private individuals. The work of the Commission, which sat from 1864 to 1868, was most comprehensive, obtaining its evidence from a wide spectrum of witnesses, from those connected with schools, from representatives of the professions and from those connected with school examinations – the Society of Arts, the College of Preceptors, the Universities of Oxford and Cambridge Local Examinations, and the University of London Matriculation Board. The Commissioners concentrated their attention on three leading subjects of instruction – language, mathematics and science.

The Schools Inquiry Commission which examined no fewer than 782 grammar schools and 2175 endowed elementary schools, presented its report in 1868.

The Commission reported that of the large number of schools inspected 'in very few cases is any effective instruction given in mathematics, French or natural science'. By far the majority of the schools, though not all, gave no better education than that of an ordinary National school. The Commission found that whilst many of the schools were not shackled by outworn statutes few of these schools had ventured beyond tradition and included science in their curricula. As late as 1868 few of the endowed schools had started teaching science.

This was true right across the country. Commissioners reporting for example in the counties of Shropshire, Worcestershire, Monmouthshire, Norfolk, Northumberland, Devon, Somerset and the West Riding of Yorkshire reported that, with a few exceptions, in the grammar and endowed schools science was hardly taught at all.

Some of the large endowed schools had begun some science teaching but its value and effect seems to have been small. At King Edward's School, Birmingham, a special department had been proposed for teaching mathematics and physics, but, as in most schools, the implementation had been impeded by 'the large machinery provided for a classical education'.[1] Tonbridge School had started classes in physics but received so little encouragement that the scheme was dropped. In most of the large endowed schools and also in the smaller grammar schools, no science was taught. This led the Commissioners to conclude that 'natural science was practically excluded from the education of the higher classes in England.'

The Commission noted that few endowed schools prepared boys for the examinations of the University of London or for the recently instituted Local Examinations of Oxford and Cambridge. The Commission found that 550 so-called grammar schools sent no boys to Oxford or Cambridge and of the 80 or 90 schools which did, only 40 sent three or more boys every year.

> Schools giving a fair general education, but laying special stress on such mechanical and physical sciences as shall best assist the scholars who are intended for manufacturing or mechanical pursuits, hardly exist. The Bristol Trade School is the most noticeable instance which has been brought before us.[2]

Of those schools that did submit candidates for external examinations the teaching left much to be desired, as the Rev. B. Price, Sedleian Professor of Natural Philosophy at Oxford and a Delegate of the Oxford Local Examinations Board, stated in his evidence to the Commission (March 1865):

> One year I examined in natural philosophy; that is, in electricity, magnetism, mechanics, and hydrostatics. The information shown by the boys as to the description of instruments was tolerably good; but when we asked for explanations of laws, which expressed physical facts, with precision and exactness, the boys for the most part completely failed. They seemed not to have gone through the mental training and discipline which enables them to appreciate, say, the laws indicated by Attwood's machine – the laws of falling bodies. They could explain an electrical machine, till you came to ask them what the rationale was, and then they completely failed. I think that has been the character of the examination in natural philosophy. The defect seems to be that they were not used to handling the instruments. There are no collections [of apparatus]; and it is one of the greatest defects in teaching natural philosophy in these middle-class schools. You cannot teach it by book; you cannot teach it efficiently by the eye, and merely seeing experiments. They must handle and use the instruments. The boys could not tell the way in which the barometer was used; in which the vernier is used, and they could not tell the laws of falling bodies. They could not tell you the principle. They would explain a pump, but they could not tell you what the weight of the atmosphere was, and they could not tell you why mercury rose to 30 inches more or less, or why water rose to 33 feet more or less.

Private and Proprietary Schools

Mr Joshua Fitch, one of the Commissioners, wrote of the denominational proprietary schools that 'the curriculum of instruction in all these schools differs in one important respect from that of the ordinary grammar school', the important difference being that 'English, geography, history and physical science receives much attention ... fancy classics, as they are called, are discarded.' He goes on to point out that 'since the upper boys are frequently destined for the University of London, the Matriculation Examination of that University furnishes the scheme of instruction'. As we saw earlier this examination demanded some acquaintance with science. Writing about Bootham School in York, which was maintained by the Society of Friends, Fitch reported that 'the provision made for the systematic study of natural science is more ample than in any school I ever visited.'[3]

The most famous of these private schools was probably Queenswood College, in Hampshire, built in 1842. It contained a well-equipped laboratory in which the boys were taught chemistry, physics and botany. Sir Edward Frankland, who had taught chemistry there, described Queenswood as the first English school to introduce the 'practical or laboratory teaching of science'.[4] His colleague at the time, teaching physics and mathematics there, was John Tyndall, who was later to succeed Faraday as Professor at the Royal Institution.

Recommendations of the Commission

The Commissioners made reference to a letter by Dr Lyon Playfair, in which he stated that the Industrial Exhibition in 1867 provided evidence of a decline in English industrial and manufacturing skills as compared to other nations and that this decline was due to a want of technical education in England. Other learned men were asked to comment on Playfair's statement. John Tyndall FRS, Professor of Physics, and Edward Frankland FRS, Professor of Chemistry, both at the Royal School of Mines, agreed with Playfair. However, it would be wrong to infer that this state of affairs was due solely to general apathy towards science. Many believed that the construction of 'an orderly and comprehensive system of science instruction' was the vital missing element.[5] Professor Acland, FRS, Regius Professor of Medicine at Oxford University, also stated this in his evidence in May 1865.

> Until we have settled exactly what amount of these subjects can be learned at a given age, and how they can best be taught, and until manuals have been written on a reasonable scale and most skilfully devised by the best minds, this question will always be in confusion.[6]

The Commission reported that the cause of science was poorly served by poor teaching arising from lack of competent teachers, lack of textbooks and lack of apparatus and facilities.

> The cause of natural science probably suffers as much from indifferent teaching as from absolute exclusion from the schools. To teach it well it is indispensable that the teacher should be as thoroughly conversant with his subject and as energetic and apt at his work as a master in any other department of knowledge. It is also desirable that he should have sufficient apparatus for illustrating his lessons, and dexterity in the use of it. At present it not uncommonly happens that natural science, accepted as a necessity, is delegated to some master of no great mark, whose task it is to get up as much information about it as may be supposed sufficient to comply with external demand. This master, besides being wanting in all but the most superficial acquaintance with his subject, is often ill supplied with apparatus, as well as deficient in skill in manipulation. Such teaching must of necessity be lifeless, unintelligent, and fruitless. In other cases it is thought that all the demands of natural science may be met by engaging an occasional lecturer to deliver a few popular lectures. Such lectures, illustrated by

startling experiments, may stir up a momentary excitement among boys, but that they can have any permanent educational value is not to be expected. Sometimes, again, it appears to be prosecuted with success in the lower forms, and then dropped altogether in the highest, simply because other subjects are better rewarded at the Universities. We cannot wonder that when it is treated in this way it should be pronounced superficial and incapable of disciplining the mind.

Men of genius, such as Faraday, had shown how science could be taught practically to young children, as he had done in his lectures at the Royal Institution. It was for men such as these to lead the way in developing experimental courses for young people.

In 1867 the British Association for the Advancement of Science presented its report 'On the Best Means of Promoting Scientific Education in Schools' to the Taunton Commission which was then sitting. The report influenced the Commissioners, though they did not accept the recommendations in full nor commit themselves to any sweeping advocacy of science teaching. They put the case presented by eminent authorities for the claims of natural science to a prominent place in the school curriculum. The witnesses before the Commission had generally agreed that experimental physics and chemistry were most useful subjects for schools, recognizing that where they had really been tried the teaching of science had been most successful. The Commissioners were convinced that, if properly taught, science could discipline the mind just as well as any other subject.

The Commissioners point out that no scheme of education could be considered complete if it omits so important a subject as science. They recommended that in the curriculum of all secondary schools three leading subjects should be used as the chief instruments for disciplining the mind – language, mathematics and science.

Their report discussed what topics of science should be taught and at what stage. They recommended a start with simple topics of physical geography, on the grounds that no apparatus was required apart from maps, then simple descriptive botany and zoology. But when children reached the age when they could appreciate scientific precision then an elementary knowledge (as against a *superficial* knowledge) of experimental physics and chemistry was to be taught. The Commissioners pointed out that there was no reason why, in the more traditional schools, classics and mathematics should fear the encroachment of science; indeed, study of these subjects could even gain from study of science.

The Schools Inquiry Commission recommended the following:

1. There should be a *central* (State) authority to control all secondary education, assisted by regional authorities.
2. Schools should be organized on three levels, depending on their purpose, their curricula and the age at which their pupils left school – 14 to 15, 16 to 17, or 18 to 19 years old. These grades corresponded roughly to the social groups of the pupils.

Schools of the *first* grade (leaving age 18) would prepare boys for the universities and so would make the classics the staple of their teaching. 'But among them it would seem expedient that some of them should be semi-classical and replace the study of Greek by more instruction in modern languages, mathematics or in natural science'.

In the *second* grade of school, with a leaving age of 16, mainly for persons of more 'straitened' means, those (grammar) schools would provide a mixed general education; often it would be worthwhile to lay great stress on practical mechanics and other branches of science as was done in the 'realschule' of Germany. These schools would be patronized largely by the mercantile class. But the most urgent educational need of the country was for good schools for the artisans, for the sons of small tradesmen. These would be schools of the *third* grade, carrying education up to 14 or 15 years old. Such schools would teach English, a language, mathematics and science.

W. Hall comments that considering all the evidence the Schools Inquiry Commission received it made remarkably little effort to further the real progress of science teaching in schools.[7] The vital need for practical work for which so many influential people argued seemed too vast for the Commissioners to accept.

The Devonshire Commission on Scientific Instruction and the Advancement of Science

This Commission sat from 1871 to 1875. The Commission's sixth report, published in 1875, dealt with scientific education – or the lack of it – in schools. The Commissioners made an exhaustive and detailed survey of schools and examined a large number of witnesses to get their views on what science should be taught, how it should be taught, at what age, and so on. Questionnaires were sent to 205 schools with annual endowments of more than £200. The Commissioners reported that

> in spite of the concurrence of authority in favour of the introduction of Science into schools, it appears that not one half of the 128 Endowed Schools from which returns have been received have even made an attempt to introduce it; and, of these, only 13 have a Laboratory, only 18 have limited apparatus for teaching science and only 10 give as much as four hours a week to these subjects.
>
> The evidence placed before us conclusively proves that in our Public and Endowed Schools, science is as yet very far from receiving the attention to which, in our opinion, it is entitled ...
>
> Among the Higher Grade Schools, only two attach equal weight to it as Classics and Mathematics in the School Examination; only 13 give any weight at all.[8]

Among the excuses given to the Commissioners for this neglect of science were: (a) the lack of funds, (b) the uncertainty as to the educational value of science, and (c) the difficulty of finding time for a new study in an already over-crowded curriculum.

1. Lack of funds

The headmaster of Loughborough Grammar School was typical when he stated: 'we have no Science taught in this school at all, not for want of *will* but for want of *means*. Science would have been introduced long ago but for £. s. d.' The Commissioners replied as follows:

> The absence of funds, no doubt places very serious difficulty in the way of some of the smaller foundations; but experience shows that satisfactory results in science teaching may be produced at a very small expense ...
>
> There is a very widely prevalent idea that science teaching is very expensive, and there is very little doubt that this idea has sufficed to keep science out of a good many schools in which it otherwise would have been at the present moment in full work. The idea has arisen from the considerations, first, that, in science, unlike other subjects, the mere book was useless; secondly, that special buildings and expensive apparatus were, therefore, of the first necessity; and thirdly, that, this being so, a special teacher, who, as a specialist, would require a higher rate of remuneration than the teacher of the more commonly taught subjects, was also essential.
>
> Let us first consider the question of buildings and apparatus. While it is perfectly true that a great sum of money may be sunk in stones, bricks, and mortar, in building a magnificent laboratory or observatory, it is equally true that science teaching of a high order may go on quite independently of such outlay or adornment.

The Commissioners referred to the way in which the Rev. W. Tuckwell of Taunton School had introduced science into his school with very limited funds. In a paper on *The Obstacles to Science Teaching in Schools* which he had read before the British Association in 1871, Tuckwell pointed out that it was ingenuity, not funds, that were lacking.

> It ought to be more widely known for how very small a sum sufficient apparatus can be obtained to teach natural history and experimental science. A laboratory can be fitted up for 20 boys at a cost of little more than £20, electrometers, telescopes, polariscopes, models of pumps, and pulleys, can be made, after a little instruction, by the boys themselves, who will learn in their construction far more of the principles which they involve than could ever be instilled into their minds by the choicest produce of the shops.
>
> At Manchester Grammar School I saw several polariscopes constructed at home by day-boys of 13 years of age, some of them actually having a metallic framework. One boy had melted up an old teapot for materials.
>
> From Clifton the evidence is to the same effect. ... elementary instruction in science can be carried on at a much cheaper rate than is generally supposed, provided that the teacher will take the trouble to think for himself, and expend a little time in preparing apparatus from simple materials.[9]

Linked with the question of funds was that of the costs of qualified science teachers, if they could be found – of men trained in the new subject and yet of appropriate social standing for work in public schools. The headmaster of Wakefield Grammar School, in his reply to the Commission's questionnaire, lamented that 'he has apparatus but no funds to find teachers'.[10]

The only science teachers available at the time were those who had passed the examinations of the Science and Art Department, and were generally of lower social standing. Consequently, as Turner points out, 'in the rare cases when a science specialist was appointed, he had an uncomfortable time in the Common Room'.[11]

2. Uncertainty as to the educational value of science

Science teaching was still opposed on educational grounds. The headmasters and governors of schools had been brought up in the classical tradition; they looked on science as having only utilitarian value, a cold-blooded intellectual feat with no beneficial effect on character. Like Blake, they felt science was making the rainbow cold; like Carlyle, they thought science was destroying the mystery of the universe. The utilitarian association of science with 'inferior' trades and manufacturers blinded parents to its cultural value.

A correspondent in *Nature* (23 March 1871) wrote: 'All that parents wish is that their sons should receive a classical and general education; but any meddling with Science I can assure you some of them look upon as simply an amusement of the master's, obtained at the expense of their son's time.'

The same correspondent stated that the cause of science teaching would be greatly advanced if some master hand, such as Professor Tyndall, would offer advice to bewildered teachers on *what* to teach and *how* to teach it.

> We have heard enough about the want of it [science]; we want now to be told by competent authority how the want may be best supplied. And I am able to say from my own personal knowledge that there are heads of schools quite convinced of the importance of the subject, but utterly ignorant how to set about remedying it.

At the time no uniform opinion had been reached on what science should be taught. As Archer points out, 'all the science specialists could agree in urging that science should be taught, but unanimity would have ceased the moment they tried to determine what science.'[12]

There was the uncertainty as to what kind of career science specialization could offer. The rewards for success in Classics was so great that a boy would hesitate before endangering his progress in them. The belief among parents was that science did not pay. There was also the unsettled state of opinion as to what was the real aim of scientific study; the instinctive feeling of boys was that, as presently taught, science was not equal to Latin or Greek for educational purposes. Many headmasters divided their schools into two sides – classical and modern – the latter giving instruction in science. Into the modern sides the dullards were often placed, and then their failure in later life was brought as evidence of the futility of science education.

The universities play a crucial role in what is done or not done in schools. The status of science in the universities would inevitably effect its introduction and teaching in the schools. The Head of University College School commented:

Every improvement the Universities introduce into their examinations must be attended with improvement in our teaching, e.g. if the examinations of the University of London became partly practical in all departments of science, as they are, for instance, at University College in the case of botany, classes under instruction would be stimulated to much greater interest in, and attention to, the details of arrangement and manipulation of apparatus; and the substitution of practical instruction for teaching by the exhibition of experiments only must soon follow.[13]

The school authorities suggested ways and means which, in their opinion, would be helpful changes on the part of the universities. Many headmasters were of the opinion that universities could help the cause of science in schools by offering more scholarships and exhibitions, since at this time there was a marked disparity between those given to science and those to other subjects. As the report of the British Association on *The Best Means of Promoting Science* (1867) put it:

At present public opinion in the University [Cambridge] does not reckon scientific distinction as on a par with mathematical or classical; hence the progress of the subject seems enclosed in this inevitable circle – the ablest men do not study natural science because no rewards are given for it, and no rewards are given for it because the ablest men do not study it.[14]

The headmasters suggested the universities should adopt special courses on the art of teaching, similar to these held in the School of Science at South Kensington. They also urged the inclusion of science in the entrance and matriculation examinations.

3. Over-crowded curriculum

The existing curriculum was supposedly over-crowded and headmasters were reluctant to introduce fresh studies. However, the Commissioners were not impressed with this argument.

The third of the difficulties most frequently urged is want of Time. While we cannot deny the reality of this difficulty, it seems to us to offer no justification whatever for the total or almost total exclusion from Education of any great branch of Human Knowledge. The difficulty is one which can only be met by carefully economizing time, by employing the best methods of teaching, and by discarding superfluous subjects of study. To meet it by making education one-sided and incomplete, cannot be for the interest of the pupil. The number of hours of study in our Public Schools may be taken at not less than 35 per week, including in the estimate the number of hours on an average employed in preparation. Now, if six hours per week be devoted to Science, and if we suppose six hours also to be given to Mathematics, there would still remain at least 23 hours a week for the study of Language and other subjects.

We are not prepared to admit that the classical Scholarship of the Pupil would by the close of his school career, have suffered in consequence of the subtraction of the 12 hours which we have assigned to Mathematics and Science;

since we believe that the influence of Instruction in Natural Science on the development of his intellectual powers might be such as to promote his success in Classical Learning.[15]

Thus the Commissioners found none of the difficulties mentioned insuperable.

Teaching methods

In 1874 a Committee on the Teaching of Physics in Schools had reported to the British Association for the Advancement of Science. It recommended that the first teaching of all branches of physics should be of an experimental nature, the experiments being done by the pupils themselves and not by the teacher. Though textbooks might serve a purpose, they should be merely for reviewing work already covered by experiment, demonstration or lecture and not for preparing subsequent lessons. It recommended that in schools the teaching of physics should begin with mechanics, including hydrostatics and mechanics, all in a purely experimental manner, to be followed by heat, optics and lastly magnetism and electricity. If time permitted elementary astronomy could be included.

The Devonshire Commissioners concurred that an effective grasp of the experimental sciences could not be acquired from textbooks alone; no science teaching could be regarded as satisfactory unless the learner became familiar with the methods of experiment and observation by practising them in the laboratory. They stressed the value of laboratory work. They suggested that the study of science, more than any other subject, offered tremendous scope for the 'development and training of the mental faculties'.

In 1872 Professor G. Carey Foster, FRS, Professor of Physics at University College, London, had written an excellent article 'On the Teaching of Physics'. This was published in the *Educational Times* of 1 February 1872. In its wisdom, the Devonshire Commission saw fit to reproduce the article in its Report. Extracts from this article are given in Appendix 6. Its rationale and approach are modern enough that any physics teacher of the 1990s would benefit from reading it.

To help teachers who had to take large classes of students in practical work the Commission drew attention to the method used by Professor Pickering at Massachusetts Institute of Technology, in Boston, USA.[16] This is basically the method still used in undergraduate physics laboratories, in which pairs of students perform different experiments at the same time, the supervisor moving from group to group.

To assist inexperienced teachers who were not too sure what physics to teach nor how to teach it, the Commissioners appended a 'Sketch of Experiments' covering sound, heat, light, frictional electricity, voltaic electricity and magnetism, that could be carried out. To assist these teachers further the Commissioners also appended a 'Catalogue of Apparatus' with prices.[17]

Conclusions

The Devonshire Commission concluded as follows:

> Considering the increasing importance of Science to the Material Interests of the Country, we cannot but regard the almost total exclusion [of science] from the training of the upper and middle classes as little less than a national misfortune. We, therefore, recommend that in all Public and Endowed Schools, a substantial portion of the time allotted to study, should, throughout the School Course be devoted to Natural Science; and we are of the opinion that not less than six hours a week on the average should be appropriated for this purpose ... That in all General School Examinations not less than one sixth of the marks be allotted to Natural Science ... The Special Commissioners have found it necessary to insist that there should be at least one Science Master for every 200 boys.

In addition, the Commissioners appealed to the Universities to give the schools the incentives they needed to embark upon science instruction.

As a result of these recommendations there was a rapid increase of science teaching in the schools. Gradually headmasters came to realize that science had come to stay. In the closing decades of the century the smaller grammar schools came into line with the large schools and built laboratories and appointed science masters. Science classes or Schools of Science were opened which devoted 18 out of 24 hours to science and mathematics. The Bryce Commission of 1895 complained that the curriculum in some secondary schools was too much dominated by science and that, in consequence, the literary subjects were being neglected!

Chapter Twelve

External Examinations

This chapter considers the role of the University of London Matriculation and BA/BSc Examinations and the Oxford and Cambridge Examining Board in establishing physics in the school examination system.

The University of London

In 1836 a Royal Charter established a federal university of London, consisting of University College, London and King's College. In 1837 this University of London was chartered as an examining, degree-granting body. The Senate included such eminent scientists as Faraday, Dalton, George Airy and Dr Neill Arnott whose *Elements of Physics*, published in 1827, became a widely used textbook of physics for the next 50 years. Not surprisingly the new university gave science an important place in the curriculum.

Matriculation Examination

The matriculation examination of the University of London has had a great influence on the curriculum of secondary schools and colleges. From its beginning the University has always been severe in its demands. Candidates had to pass in all the following subjects:

1. Latin.
2. Any two of Greek, French, German, Sanskrit or Arabic.
3. English language, history and geography.
4. Mathematics.
5. Natural philosophy (mechanics, hydrostatics, hydraulics and pneumatics, acoustics, optics and heat).
6. Chemistry (including heat) and natural history.

Honours were obtained by passing extra papers in mathematics and physics. These regulations remained in force until the 1880s, when magnetism and electricity were included. In 1885 candidates still had to satisfy conditions 1, 2, 3 and 4, but 5 and 6 were amended as follows:

5. Mechanics
6. One of Chemistry, Heat and Light, Magnetism and Electricity, Botany.

In 1899 candidates had to take two papers in general elementary science and one paper in any one of elementary mechanics, heat and light, magnetism and electricity. The compulsory papers in general elementary science consisted of questions on statics, dynamics and hydrostatics, heat, light and chemistry. The regulations stipulated that 'the subjects shall be treated wherever possible from an experimental point of view'. Thus till the close of the century, some knowledge of physics and chemistry was required by all London Matriculation candidates.

During 1838–42 candidates from 28 schools were successful in the examination. By 1851 another 17 schools had successful candidates. Up to 1851, 1675 candidates in all had taken the examination, 1493 being successful.

In his evidence to the Public School Commission in November, 1862, Dr W.B. Carpenter, FRS, who was Registrar of the University of London, pointed out that 'the questions which are asked are not merely what are the laws, but questions involving the application of those laws to cases which may not have been within the knowledge of the student'. Carpenter explained that these subjects were not to be treated mathematically. When asked by Lord Clarendon how the line could be drawn between the facts and the mathematical demonstration of them, Carpenter replied:

> The composition and resolution of forces, for example, treated mathematically, requires an advanced knowledge of mathematics; but the general fact that a body acted on by two forces in different directions will move in a diagonal line which expresses the result of those two forces, and the mode in which that fact is applied in the common occurrences of life, such as, for instance, rowing a boat across the stream with a tidal current, are subjects of examination with which the candidate is expected to be conversant. So again, with the pressure of liquids and gases, and the knowledge of what specific gravity means and of how specific gravity may be ascertained; with the principle of the siphon, the common pump, the forcing pump, and the air pump. In acoustics he should be able to explain the nature of sound, and to answer all the elementary questions in regard to echoes and the rate of the transmission of sound, but without any mathematical treatment of those subjects. In optics he is expected to understand the laws of reflection and refraction, and the formation of images by simple lenses. Questions are asked to test the candidate's knowledge of those laws, and his power of applying them, but not going into the higher departments of optics. Under the head of Chemistry, the candidate is required to have studied the phenomena and laws of heat, the sources of heat, the general effect of the expansion of bodies, the action of the thermometer, and the relation between the different scales in common use. Such questions are frequently asked, as, for instance, 'convert 60° of Fahrenheit into the corresponding degree of a Reaumur or a Centigrade thermometer.' This is useful to every person who reads foreign scientific books; as these express temperatures in degrees of the Centigrade scale, with the exception of those that give them in Reaumur. Then there is the difference between the temperature and quantity of heat, the doctrines of specific and latent heat, and the various modes of measuring heat, the action of heat on the condition of bodies producing liquefaction, ebullition, and evaporation, the doctrines of conduction, convection, and radiation of heat; all these being subjects a knowledge of which is applicable in all the familiar concerns of life.[1]

BA and BSc Examination

The London BA examination was introduced at the same time as the London Matriculation. The syllabus for the BA pass degree in 1839 and the regulations and examination papers for the London BA Honours examination of 1839 are given in Appendix 7.

In 1850 a new charter was given to the university and candidates were admitted to the degree examinations from affiliated colleges in various parts of the country. This system of external degrees proved a great boon to technical colleges and even university colleges before they acquired their own full university status.

In 1860 new regulations for the London BA were issued. Until 1876 the BA pass degree included a significant amount of science which was compulsory for all candidates. Mechanical and natural philosophy constituted one of the compulsory subjects. In 1881 things changed for the worse as far as science teaching was concerned – it became possible to gain a BA without offering any science at all. This drawback was compensated for by the introduction of the BSc degree. The syllabus for the Inter-BSc pass examination of 1882 is given in Appendix 8.

The Oxford and Cambridge Local Examinations

In 1857 a Committee met at Exeter to establish a system of examinations and prizes for boys to test a good general education, including the 'general principles of natural philosophy'. That year 106 candidates presented themselves, including 'two little fellows from a National School having walked 20 miles to attend'.

At the instigation of Dr Temple (headmaster of Rugby, and later an archbishop) the two universities took up the idea and created local examinations for schools, Oxford in 1857 and Cambridge in 1858. These **Local Examinations** were to meet the needs of middle-class schools. The universities would provide the guarantee of scholarship and impartiality which both parents and teachers wanted. The examinations were in two divisions – the junior for boys under 16, the senior for boys under 18.

The l858 Oxford Local Examination was held in 11 centres. Of the 401 senior candidates, only 150 were successful. The science subjects attracted very few candidates, as is seen from these figures: mathematics (308) was as popular as French (299); but natural philosophy had only 77, chemistry 76, psychology 12.

There was a remarkable growth in numbers for the Cambridge Local Examination. Whereas at the first examination, in December 1858, 297 junior and 73 senior candidates appeared, 20 years later the total number of candidates was 6435, made up of 626 senior boys, 997 senior girls, with 3329 junior boys and 1483 junior girls.

At the 1878 Cambridge Senior Examination for boys 461 took pure mathematics, 167 applied mathematics, 92 chemistry, 21 statics, 38 heat, 30 electricity, 15 zoology, and 11 botany. Considerably more senior girls took zoology (79) and botany (177) than pure mathematics (76); 962 senior girls took French. In 1873 a girl headed the list of seniors in the Oxford examination, and because there was nothing in the list to indicate her sex, she was offered an Exhibition at Worcester College, Oxford! These local examinations were one of the most important levers in raising the whole level of women's education throughout the country.

The College of Preceptors

The College of Preceptors was concerned with raising the professional standing of teachers. It was the first body to institute lectures on the science and art of education for secondary teachers.

External examinations at secondary level for both grammar schools and private schools, and for teachers, were begun by the College in 1854. These were the first external examinations to be held for middle-class schools in the country. In the examinations natural philosophy, chemistry and natural history were optional subjects and were not required for a certificate. In 1870 there were 1571 candidates for the pupils' certificates of the College.

However, although these school examinations performed a very useful function, the College lacked the prestige or authority to play a major role in the overall direction of secondary education.

Chapter Thirteen

The Heuristic Method of Science Teaching
H.E. Armstrong (1848–1937)

Professor H.E. Armstrong was the great apostle of the heuristic method of teaching. While others wrote about it he practised it in his own teaching. In *The Aims and Practice of Teaching Chemistry* (1897) he wrote: 'But so long as we merely teach children what is, and do not equip them to be – in however humble a manner – discoverers in their turn, we manifestly fail to educate them to the best advantage.'

The heuristic, or discovery, method is one in which the learner is put in the position of an investigator, the teacher so manipulating the subject matter and activities that the learner discovers facts *for himself*. Throughout his writings Armstrong states that the beginner can and must be put in the position of an original discoverer. Young children are delighted to be told they are to act as young detectives: 'children are from the outset born enquirers, and all we have to do is to develop and practically train faculties which they all possess to some extent.'

The philosophy behind Armstrong's heuristic method of teaching science is well illustrated in his *Notes on the Teaching of General Elementary Science and the Elements of Scientific Method*, which he wrote for teachers in training (1901). In this he shows how a whole series of connecting lessons can be developed logically with the children themselves doing and discovering things which lead up to simple laws and concepts of physics, such as density, Boyle's Law (without using mercury), energy as exemplified in the steam engine, and so on; and all with the very simplest of apparatus. He would introduce a topic with an interesting story. 'Could some of the statements in the story be true? Well, let's find out!' And the children, using their own eyes and hands, do find out. (Extracts from Armstrong's Notes are given in Appendix 9.)

Armstrong did not approve of the methods of education of the time. He disapproved of

> didactic class teaching, extravagant indulgence in the use of books, and disregard of all tools and weapons other than the pen ... The use of textbooks must be carefully avoided (at the early stage) in order that that which should be elicited by experiment is not previously known and merely demonstrated.
>
> I want to see the workshop method introduced – i.e. specific tasks set in the form of *problems* which each child is to work out experimentally, the teacher merely acting as foreman. When experimental work is done in this way, it tends to develop habits of independence, children become observant and thoughtful, their imagination is called into play; if properly led they become exact and honest workers; instead of becoming parrots, they learn to think and act for themselves. It cannot be too strongly insisted that the object in

> view – the training of the faculties – requires, not that information should be *imparted* but that information should be *gained* by personal observation and experimenting.
>
> The great object of the teacher of 'science' should be to teach the *art of experimenting*... To experiment is to search out by trial. The essential first step in an experiment is to have a clear conception of the nature of the quests in which it is proposed to engage. The motives with which each experiment is made must be clearly understood ... An experiment, like an act, 'hath three branches' – to conceive, to do, to utilize; a clearly defined motive must underlie it; it must be properly executed; the result must be interpreted and applied. It is only when the motive is clearly written out that it is clearly understood – that the meaning or intention of the experiment is clearly grasped.[1]

Armstrong was a great advocate of the need to write down the several aspects of any experiment clearly and neatly.

> The writing out causes an understanding of the motive – makes it clear that there is an object in view. Then, as the work is done, the exact way in which it is done is recorded there and then; and drawings are made so that there may be no doubt as to what were the arrangements adopted. Again, in writing out this part of the account, every step taken has to be explained and justified. In the third place, the results obtained are carefully recorded. In the fourth, the deductions to be drawn are pointed out. And in the last, the ways in which the observations make it desirable to try some further experiment are fully discussed.
>
> In this way, the scholar is not only led to experiment with a set purpose and to acquire the habits of observing and reasoning logically from observation but each experiment becomes a most valuable lesson in composition and therefore in language. The teacher of the future must be guide, philosopher and friend to the taught – not a mere trainer of parrots. The teacher must cut himself adrift from codes, and become a self-acting, reasoning being not a mere automaton wound up once for all at a training college.[2]

Armstrong disapproved of examinations, such as those organized by the Science and Art Department. He was not prepared 'to worship at the examination shrine to earn the unholy money bribe called payment by results'. Furthermore 'the extent of ground to be gone over is so great as to make cram an absolute necessity'.

On 4 May 1933, Armstrong, then an old man, gave a Huxley Memorial Lecture at the Imperial College of Science and Technology in London. In this he gives some revealing insights not only into his own education but of these personalities who were to influence him greatly as his ideas on educational methods crystallized.

Armstrong was little impressed by Huxley as a teacher. Huxley's fame as a master of education rests mainly on his writings and public addresses, not upon his work as a teacher.

> More than 65 years ago, a young beginner, as a student, in 1886–7, I sat, if not exactly at his feet, in face of his blackboard and watched him embroider it most exquisitely with chalks of varied hue: the while he talked like a book: with absolute precision, in chosen words, so easily that we were hypnotized by his

basilisk artistry into the absurd belief that we were learning: in fact we were just being told, allowed to have no doubts, with no time to think! He was a marvellous exponent – therefore a bad teacher, as are all who are eloquent. This was at the School of Mines in Jermyn Street. The course was purely formal and descriptive, without living illustration.

> In his lectures, Huxley was the letter-perfect exponent of his facts: master of a fine logic but encased in hard bones – essentially a didactic teacher – with a marked tendency to pontificate.[3]

Armstrong went on to describe the science teaching of his time. 'They are taught didactically and learn by rote: their experiments are merely verifications or demonstrations in proof of assertions: the art of *discovery* is seldom, if ever, taught either in University or school.[4]

There was however one redeeming star – John Tyndall, Professor of the Royal Institution.

> *Per contra* attending Tyndall's marvellously illustrated lectures on Physics, also at Jermyn St, I was enraptured. His Icelandic Strokur, spouting up to the ceiling ever lives in my mind's eye. Everything in *Heat as a Mode of Motion*, everything in his *Sound*, was put before us: we really knew what wave motion was. Tyndall's science may have been limp at times – he was no mathematician but as a teacher he was worth all the Professors of the Royal College of Science and Royal School of Mines put together and no other actor on the Royal Institution stage has ever surpassed him as a popular lecturer.[5]

Early in his career Armstrong lectured in chemistry at St Bartholomew's Hospital, where he was astonished at the inability of his medical students to interpret simple experimental results and their general lack of any critical stance. He believed that only the teaching in the schools could produce such a sorry state of affairs. It was in the course of a lawsuit in 1880 that Armstrong was fully convinced of the importance, not so much of facts, but of method.

> The display of judicial method, 'the stringent examination and cross-examination of every particular' came to me as the 'acme of scientific thought' ... I thus at last realised that knowledge (science) was one thing; the methodical logical use of knowledge (scientific method) another; the judicial was something far higher.[6]

The main criterion was *to do* so as *to understand*. What was learned was of less importance than the *method* used in learning, thinking, finding out. Once the student had learned the method (experimentation) he could continually use the technique of 'finding out' and so painlessly acquire by his own activities the mass of information which the existing teacher-centred systems of education pumped into children by rote, demonstrations, or examinations. The essence of Armstrong's thinking were summarized in an article in *Nature* in January 1904:

> Two things and two things only, I think, are essential to Professor Armstrong's plan, first, that the pupils should perform experiments with their own hands,

and second, that these experiments should not be the mere confirmation of something previously learned on authority, but the means of eliciting something previously unknown or of elucidating something previously uncertain.[7]

Armstrong advocated the introduction of scientific method into the schools. Scientific method was to be a basic staple in the school curriculum, as were the three Rs. Just as all children must be taught to read,

so all children must be taught the elements of scientific method ... The solution I would venture to offer is that an honest attempt be made to teach not only the three Rs but also a fourth – Reasoning – the use of thought-power. The fourth R – Reasoning – will necessarily be taught in connection with every subject of instruction.[8]

According to Armstrong, this scientific method must be used right from the start of education.

I am satisfied that the one remedy ... is to commence experimental studies at the very earliest possible moment, so that children from the outset may learn to acquire knowledge by their own efforts, to extend infantile practice – for it is admitted that the infant learns much by experimenting – and the kindergarten system into the school, so that experimenting and observing become habits.[9]

What Armstrong argued for was 'to substitute rational for mechanical methods of teaching, in order to rouse the inquisitive tendencies which in many cases now are deadened; to make the school, not the early grave of individuality, but an attractive spot.'[10] Armstrong appreciated the difficulties of introducing the scientific method into the schools. Teachers 'should have mastered experimental method and be able to assume the attitude of the investigator'. But, as he points out, 'at present so few can appreciate what is meant by such expressions.'

Armstrong places the onus of producing 'a new race of teachers' on the universities.

Teachers such as we need will not be forthcoming unless the universities take a far broader view of the situation than they have done heretofore. It will not suffice to supplement the ordinary degree course by a year's study of pedagogics. No amount of didactic teaching will make teachers; little more than the mere tricks of the trade will be learnt in the practising school: real teachers will only arise when the training given is such as to develop thought-power and some understanding of the art of experimental inquiry.[11]

Though heurism is generally associated with the name of Armstrong he was not the first to use or to refer to a heuristic method. Armstrong himself stated

Let me here at once say that I am not responsible for the introduction of this word [heuristic], still less of the principle included under it. The method, in a sense, is as old as the hills – in fact, it is the method of nature ... Its use has been advocated over and over again. But it had fallen into disuse, having been almost lost sight of since literary methods have secured the mastery in schools ... The heuristic method is nothing more than the method of gaining knowledge by experience – by inquiry, by observation, by trial – reduced to system and formalised.[12]

Armstrong points out that over a century ago one of England's greatest discoverers, Joseph Priestley, had recommended the heuristic method. Priestley wrote:

I would observe that, if we wish to lay a good foundation for a philosophical taste and philosophical pursuits, persons should be accustomed to the sight of experiments and processes in early life. They should more especially be early initiated in the theory and practice of investigation, by which many of the old discoveries may be made to be really their own – on which account they will be much more valued by them.[13]

Without calling their methods 'heuristic', there were teachers who were using the technique in their teaching. Thus the Rev. J. Bryce, in his evidence to the School Inquiry Commission (May 1866), gives an example of some inspired physics teaching along heuristic lines at Belfast Academy.

One great advantage of the study of physical science is that, when properly taught, it interests boys in intellectual pursuits generally. For instance: Newton's great discovery, the identity of the power which retains the moon in her orbit with terrestrial gravity, was being explained to a class of from 12 to 18 boys. The teacher did not tell them the result: he enumerated the phenomena by which Newton arrived at it, taking care to present them in the order most likely to suggest it. As fact after fact was marshalled before them, they became eager and excited more and more, for they saw that something new and great was coming; and when at last the array of phenomena was complete, and the magnificent conclusion burst upon their sight, the whole class started from their seats with a scream of delight. They were conscious that they had gone through the very same mental operation as that great man had gone through. The consciousness of fellowship with so great a mind was an elevating thing, and gave them a delight in intellectual pursuits. An unusual proportion of those boys who passed through the Belfast Academy during the 20 years that I was able to have natural and physical science taught on those principles, have, as men, been distinguished and successful; and they owe it, I am convinced, in a large degree to the taste for intellectual pursuits thus formed.[14]

Joseph Payne, who was the proprietor of a large boarding school at Denmark Hill, London, and Professor of the Science and Art of Education to the College of Preceptors, stated in an essay on the 'True Foundation of Science Teaching' that

The learner is a discoverer or investigator engaged in interrogating the concrete matter before him and the teacher is a superintendent or director of the learner's process, pointing out the problems to be solved, concentrating the learner's attention upon, varying the points of view, suggesting experiments, enquiring what they result in; converting even errors and mistakes into means of increased power, bringing back the old to interpret the new, the known to interpret the unknown, requiring an exact record of results arrived at – in short, exercising all the powers of the learner's mind upon the matter in hand in order to make him an accurate observer and experimenter and to train him in the method of investigation. [15]

Payne refers to such 'heuristic' teaching carried out by Dean Dawes, by the Bristol Trade School, and to

the invaluable lessons given to the imaginary Harry and Lucy by Miss Edgeworth ... I have no hesitation in declaring that they one and all acted mainly on the principle that true Science-teaching consists in bringing the pupil's mind into direct contact with the facts – in getting him to investigate, discover and invent for himself.

For Payne, 'the true foundation of physical Science lies in the knowledge of physical facts gained at first hand by observation and experiment, to be made by the learner himself'.[16]

Canon J.M. Wilson, science master and later headmaster of Rugby School, criticized current modes of science teaching, as in the lecture method, where the pupil plays a largely passive role.

The scientific lecture in which results which cast centuries of thought are expounded to the pupil in the course of a few minutes fails to exercise the active faculties required for an investigation in any matter ... These may lie dormant in the class while the most learned lecturer experiments with facility and explains with clearness ...

Training to think must be first and foremost the object. The teacher ought to make the pupils teach themselves, by thinking out the subject with them, by taking up their suggestions and illustrations, by criticising and hunting them down, by starting them on a fresh scent when they are at fault, by reminding them of some familiar fact which they had overlooked, and so eliciting out of the chaos of vague notions that are afloat something of order and concatenation and interest before the key to the mystery is given, even if after all it has to be given.[17]

Wilson placed great stress on the learner himself taking part in investigations.

Dr Kemshead, science teacher at Dulwich College, was another to favour the heuristic method of science teaching.

To develop scientific habits of thought – the scientific mind, the teaching must be of a totally different nature. In order to get the fullest benefit from a scientific education, the teacher should endeavour to bring his pupil face to face with the great problems of nature, as though he were the first discoverer. He should encourage him from the first to record accurately all his experiments, the object he had in view in making them, the results even when they have failed, and the inferences that he draws in each case, with as much vigor and exactitude as though they were to be published in the 'Philosophical Transactions'. He should, in fact, teach his pupil to face the great problems of nature as though they had never been solved before.

'To face the great problems of nature as though they had never been faced before' – 'to bring the child face to face with the great problems of nature, as though he were the first discoverer' – those weighty, pregnant and luminous expressions contain the essence of the whole question I have endeavoured to set before you. They define the attitude of the pupil in regard to his subjective process of learning, and the function of the teacher in regard to his objective process in teaching – the one being the counterpart of the other.[18]

Whilst many voices argued for more enlightened methods of teaching, it was Armstrong who not only shouted the message of heurism from the housetops, but put it into practice in his teaching.

In 1879 the City and Guilds of London Institute was formed as a technical college. Anxious to make an immediate start the College created two lectureships, one in applied chemistry and one in physics. In October of that year Armstrong was appointed to the former whilst to the latter the College appointed the electrical engineer W.E. Ayrton (1847–1908), who had just returned from teaching in Japan. Ayrton had the advantage of teaching a brand new subject – electrical engineering. His one object was to make his men practical electricians. Though the course was practical, it was too didactical. As Armstrong later pointed out:

Some of us irreverently termed it the penny-in-the-slot method. The student found himself in face of certain apparatus and the instructions he needed to carry through the verification of this or that theorem – Ohm's Law, for example.

The method had the great advantage, that the student learnt his lesson by doing instead of being merely told – but with a minimum of effort and little call to think for himself. No element of discovery was involved in the exercise.[19]

After a long struggle Armstrong managed to persuade Ayrton and the Council of the Institute to agree to have one composite building in which all the first-year subjects (mathematics, engineering drawing, elementary workshop practice, chemistry and physics – mainly electricity) were common to all students; specialization was deferred to the second year. Classes were begun at Couper Street with up to 800 day and evening students. This first English technical college – the forerunner of the polytechnics – moved into a new building in Finsbury in 1883. There was no control of the subjects by any external examinations: 'we insisted on being entirely free'.

Armstrong describes his method of teaching at the college thus:

I determined to attempt to lead students to be discoverers from the outset – to train them to use their eyes and to think for themselves, on all possible occasions ... I simply developed the age-old method of the infant and the detective in place of didactic teaching. I taught very little chemistry but a great deal of chemical method.

The essence of the system lay in first posing a question or problem – in fact, in stating the crime to be unravelled. Conjoined with this was the obligation to

> write down, as a coherent story, *there and then*, as the inquiry went on, in direct, descriptive language, readable and judicial, whatever was thought and done, whatever conclusions were deduced. The greatest nonsense is talked about the impossibility of making the student discover everything for himself – no one asks that he should or believes that he can: only that he shall learn how discoveries are made, first by children, then by ordinary detectives, then by discoverers proper ... Apparently, teachers in schools, with rare exceptions, never will: they have no imagination having never found out anything by themselves or learnt to dream.[20]

Fifty per cent of Armstrong's students who took the University of London examinations in chemistry failed. Armstrong admitted that his excessive failure rate was due to his department's emphasis on practical work as opposed to book work. But in terms of research output and honours gained by his former students his department had more prestige than any other chemistry department in London. Nevertheless such failure rates produced the unpleasant implications that he was a poor teacher.

Armstrong launched his campaign to establish science teaching along heuristic lines in a lecture 'On the Teaching of Natural Science as Part of the Ordinary School Course' to an International Conference on Education held at South Kensington in 1884. At the 1888 meeting of the British Association for the Advancement of Science he drew up a heuristic scheme for the teaching of chemistry.

The 1889 Report of the British Association which was held in Newcastle bears the imprint of Armstrong.

> It cannot be too strongly insisted that elementary physical science should be taught from the first as a branch of mental education, and not mainly as useful knowledge. In order that the study of physical science may effect this mental education, it is necessary that it should be employed to illustrate the scientific method in investigating nature, by means of observation, experiment, and reasoning with the aid of hypothesis; the learners should be put in the attitude of discoverers and should themselves be made to perform many of the experiments.[21]

At this Newcastle meeting Armstrong presented his plan for a course of elementary instruction in physical science along heuristic lines – *Suggestions for a Course of Elementary Instruction in Physical Science*. This was a detailed syllabus for students from the age of seven to sixteen. Armstrong's course was divided into six stages, each of which stressed 'doing'. The first stage – 'Object Lessons' – approached the physiography of Huxley. Once children had gained proficiency in elementary arithmetic, they were ready for stage II – 'Lessons in Measurement', which included the balance, the determination of relative densities, measurement of temperature. In stage III the physical and chemical effects of heat on common substances were studied, accompanied by the inculcation of note-taking by students.

The course would be cheap since little apparatus was needed. The 'laboratory' did not need to be elaborate. The only expensive items were a set of good balances and weights. 'Six of these, sufficient for a large class would cost about £18'. To overcome the problem of large classes Armstrong suggested ways of sharing out the practical work, for example, by sets, as A.M. Worthington was doing at Clifton College, Bristol.

Armstrong's campaign for more experimental work in the schools was aided by the passage of the Customs and Excise Tax Act of 1890 which made funds available to local authorities for technical education (see page 97). The authorities used the 'whisky money' to build new laboratories and lecture rooms in the knowledge that science teaching gained grants from South Kensington. Practical science teaching on a large scale became possible.

Not everyone agreed with Armstrong. William Ramsey (of argon fame) did not agree fully with Armstrong's heuristic notions. In three lectures to the College of Preceptors in March 1891 he criticized Armstrong's syllabus in some detail.[22]

Dr R.W. Stewart, DSc, Principal and Professor of Physics at the Hartley Institution, Southampton, also had reservations about Armstrong's heurism. He argued that in early science teaching each step must be clear and definite for the pupils. Unless pupils are told what to look for, he said, time is wasted because of the vague way pupils set to work. He considered it possible to have too much experimenting and too little theory and that much of the usual laboratory work was valueless because the pupils have insufficient theoretical knowledge.[23]

> My own experience leads me to think that simple lectures, illustrated by a few really essential experiments and diagrams, give the best results ... A practical course of this kind should be conducted by means of demonstration lectures, in which the methods of work are explained and illustrated by the teacher, and should be combined with individual Laboratory Work on the part of the scholars.[24]

A large part of the early work in physics was essentially a course in measurement, and this was characteristic of much of the physics teaching in schools during the first quarter of the 20th century. It consisted of measurements involving the use of vernier callipers, screw gauge, the balance and so on.

Whilst much of early physics teaching did generate into boring measurements of everything in sight, for Armstrong measurement was important, not as mere exercises in measuring, but as crucial parts of an inquiry: 'Physical science reposes on a basis of exact measurement, so that the introduction of experimental work into schools involves the introduction of measurement work as a matter of course.'

There were fears from school teachers that if they employed Armstrong's practical, heuristic method, the Department of Science and Art might not pay grants. They were reassured by Armstrong's assistant at Finsbury, John Castell Evans, who pointed out students could well use the heuristic method and still pass the Science and Art examinations. In fact, he and Armstrong had both been engaged in teaching a class of boys from Cowper Street Middle Class School since 1880 and though the boys carried out individual experiments and though the South Kensington syllabuses were not followed slavishly, they had obtained good results in the special physical science examinations (i.e. light, sound, heat etc.). But despite their reassurance many teachers felt that the system of payment by results made Armstrong's method impractical.

Armstrong held Saturday morning sessions on methods of teaching science for London science teachers at the Central Institution. Nearly 200 teachers attended the first two sessions, and the course was repeated over the years. One of these teachers was Hugh Gordon, an Oxford graduate, who in January 1881 had been appointed by the London School Board as peripatetic science demonstrator in East London and Tower Hamlets division, one of four districts into which London was divided. Gordon himself was later to hold training sessions on heuristic methods in a deserted rice mill in Whitechapel. Between 1891 and 1893 Gordon introduced a syllabus based on Armstrong's British Association syllabus into the 15 London Board schools under his charge. Gordon's syllabus was accepted by the Education Department in 1893; at the same time the Department of Science and Art agreed that preparation for its examinations should favour experimental work. In the same year Gordon published his *Elementary Course of Practical Science*. This was in the excellent Macmillan shilling *Science Primer* series edited by Roscoe. Gordon's book was the first of many to adopt the syllabus. The book was a manual for teachers. It indicates how classes as large as 60 to 70 could be taught practically. Experiments would be made by scholars in turn; while six pupils were experimenting at a table, the others could be engaged in writing up notes, answering questions, doing calculations. He suggested to them that the method of teaching 'should be that of *suggestion* combined with a minimum of demonstration – that of asking questions, not of answering them'. Teachers should give lessons of 40 minutes at least twice a week.

The range of work covered in the course for the Board Schools included measurements of length, area and volume, weighing, relative density, the keeping of barometer records and the usual means devised to illustrate the effects of expansion by heat.

Gordon's work was continued by Mr Mayhowe Heller, BSc, also a peripatetic teacher with the London School Board and another of Armstrong's pupils. By 1897 the Armstrong-Gordon-Heller elementary science course was in operation in more than 40 London Board Schools. The heuristic method was making considerable headway.

In 1895 the Incorporated Association of Headmasters formed a committee on science teaching to devise a science syllabus suitable for boys and girls. Armstrong was a member of that committee. The Headmasters' syllabus, which was published in 1896, was printed in considerable detail, because as Armstrong pointed out, 'so long as teachers need guidance – as the majority of those in schools do at present – it is of inestimable value in promoting sound teaching.'

The syllabus covered the measurement of (1) length, (2) area, (3) volume, (4) mass, (5) density, (6) thrust and pressure, of pull and tension, (7) force, (8) temperature, (9) quantity of heat, (10) vapour pressure, (11) force in lb, and application of these various quantities. The syllabus also covered (12) resolution of forces, (13) equilibrium of three forces, (14) equilibrium of four or more forces, (15) parallel forces, (16) centre of gravity, (17) principle of moments, levers, (18) simple machines.

The syllabus emphasized practical work by the pupils themselves. But there is a vast difference between work that is merely practical, for example, measurement for the sake of measurement, and work that is truly heuristic. One section of the syllabus begins with abstract concepts such as force, then goes on to practical 'gestures' on the parallelogram of forces using pulleys and pieces of string, then considers the conditions for equilibrium etc. and finally ends with simple machines. An approach that began with simple machines and then investigated how such machines worked would be more true to the heuristic spirit.

Despite its faults the syllabus brought in a system of teaching that was far better than the old system of lectures, with laboratory work an occasional reward for the elder pupils. Though the mere insistence on pupils' practical work in no way guaranteed heuristic learning, yet the issue of this syllabus and its adoption in schools played an important part in introducing sounder methods of science teaching in this country.

The Headmasters' syllabus was approved by the Oxford and Cambridge Local Examinations Boards, who each issued a syllabus in elementary physics and chemistry on the same lines.

Armstrong favoured a single general science subject – Nature Knowledge, he called it – instead of separate science subjects. 'As soon as any one specific subject is taught, there is at once a tendency to treat it from a specialist's point of view.'[25]

He wanted to get rid of formal science and to give broad training in scientific method. The subject of general elementary science with practical work was later approved by the Headmasters' Association and was soon introduced into the Oxford and Cambridge Local Examinations. In due course the University of London prescribed general elementary science, which included the elements of mechanics, physics and chemistry, as an obligatory subject for the matriculation examination. This strongly encouraged a less academic treatment of physics and chemistry.

In 1898 the Board of Education published Armstrong's article on 'The Heuristic Method of Teaching or the Art of Making Children Discover Things for Themselves' in its *Special Reports on Educational Subjects*. This gave further impetus to the heurism band-wagon.

By the end of the century there was considerable evidence that Armstrong's campaign was meeting with success. *Nature* magazine stated 'it is not too much to say that this [Armstrong's] scheme started a revolution which gathers strength every day'.[26] Many schools had been equipped with laboratories (there were 1165 school laboratories in Great Britain) and it had been shown that large classes and the requirements of external examinations were not incompatible with heuristic methods. A small nucleus of teachers had been trained to use the heuristic method. At the beginning of the 20th century more than 40 Board Schools in London were using a syllabus based on the one prepared by Armstrong. By 1907 the heuristic method was 'very general' in public elementary schools. By 1908 laboratory classes were almost universal in maintained secondary schools and Armstrong's method was widely used.

Chapter Fourteen

Physics Teaching in the Universities during the 19th Century

During the first half of the 19th century the universities of Oxford and Cambridge were still being criticized as conservative and decadent institutions as far as teaching science was concerned. When Baden Powell, a scientist, was appointed Savilian Professor of Geometry at Oxford in 1827, Copleston, the Provost of Oriel, advised him that it would be useless for him to lecture as he would not get an audience. Science was treated as a hobby of a few eccentric dons. As late as 1868 Huxley, in his evidence to the Taunton Commission, remarked:

> I think if the manufacturer who sent his son to Oxford or to Cambridge to take an ordinary degree expected that he would thereby secure the smallest atom or scintilla of scientific knowledge, he would be mistaken. Some time ago I was at a dinner party, where there were a number of very distinguished university men, several of them fellows of colleges; and as we were discussing this very topic, I asked whether it would be a fair thing to say that any one might have taken the highest honours at the university, and yet might never have heard that the earth went round the sun? and all the gentlemen present with one consent said, 'Yes.' That gives some notion of how far science forms a necessary part of the training at the universities.[1]

Other evidence suggests, however, that things were not quite as bad as that.

Oxford

From the beginning of the century the BA and MA examinations at Oxford included physics. In 1807 a special School of Mathematics and Physics had been set up. The candidates for the highest mathematical honours had to satisfy the examiners not only in pure mathematics but also in mixed mathematics which included mechanics, optics, hydrostatics, and astronomy.

At the start of the century the Sedleian Professor of Experimental Philosophy was Thomas Hornsby (1733–1810). Hornsby had a large collection of apparatus with which to illustrate his lectures. It was valued at £375 14s 6d – a large sum of money for those times.

In addition to the Sedleian Chair of Experimental Philosophy, a special Readership in this subject was established in 1810. The first Reader was Stephen

Rigaud FRS (1774–1839). Rigaud gave two courses of lectures every year, which comprised the principal experiments in mechanics, hydrostatics, optics, electricity and magnetism. When the University Press moved out of the Clarendon Building in 1831, two rooms were fitted according to Rigaud's wishes for the natural and experimental philosophy classes. These two rooms formed the Clarendon Laboratory. One room was furnished with apparatus, including a model of a steam engine, and could accommodate 70 students. The other room was 'for lectures not requiring any fixed apparatus'.

Rigaud was succeeded in 1839 by the Rev. Robert Walker, MA of Wadham College, who, in 1830, published *The Elements of the Theory of Mechanics*, an advanced textbook of statics and dynamics. Describing lectures in the Clarendon laboratory W.Tuckwell writes in his *Reminiscences of Oxford* (1900) that 'it was there that "Cheery Mr Walker" ... constructed and exploded gases, laid bare the viscera of pumps, and steam engines, forced mercury through wood blocks in a vacuum, manipulated galvanic batteries, magic lanterns, air guns.'[2]

Physics was taught not only by the Professor and the Reader of Experimental Philosophy. Several branches of physics, such as heat, light and electricity, were also taught by the Chemistry Professors. John Kidd was the first occupant of the Chair of Chemistry established in 1803. He lectured on light, electricity and heat in the Ashmolean Museum. Kidd was succeeded in 1822 by Dr Charles Daubeny (1795–1867) who also lectured on these branches of physics. The apparatus then existing in the Chemistry Laboratory (in the Ashmolean) consisted of air-pumps, pendulum, lenses, microscope, electrical machines, Leyden jars, electrometers, electrophorus, Faraday's magnetic apparatus, Ampère's magnetic apparatus, apparatus for decomposing water, galvanic batteries, pairs of four inch galvanic plates, eudiometers, magnetic needle on stand, Leslie's differential thermometer, pyrometer, hydrometer, barometer, and flasks and screws for weighing air.[3]

The textbooks in use at the time for the mathematical sciences were Robert Walker's *Hydrostatics*; Wood's or Coddington's *Optics*; Maddy's or Brinkley's *Astronomy*.

Several reforms were proposed to make the study of the physical sciences more widespread. The Rev. Baden Powell, the Savilian Professor of Geometry, one of the most able natural philosophers of the day, argued in 1832 that:

> instead of fewer Honours candidates taking an increasingly difficult paper in the mathematical sciences, it would be preferable to include a wider range of the physical sciences. This would encourage the *many* to pay some attention to these subjects, instead of leading only the *few* to pursue them to a high degree; so as to make an acquaintance with at least the principles of Science become as *essential a part of a liberal education*, as a knowledge of the classics is at present considered to be.[4]

The Head of Christ Church was already putting this into practice. A course on experimental philosophy, followed by satisfactory completion of an examination, was mandatory on all undergraduates at the college. In 1833 a proposal to 'render imperative some branch of Natural Philosophy' as a qualification in all candidates for the BA degree was vetoed by the Vice-Chancellor. In 1840 Daubeny, the

Professor of Chemistry, again raised the question 'whether an examination on the leading truths of Physics and Chemistry ought not to be gone through by every candidate for the degree of BA.'[5]

There were, of course, objections to such proposals. Sir William Hamilton, for instance, argued that the new subjects – such as science – 'are essentially easy; requiring comparatively little talent for their promotion, and only the most ordinary capacity for their acquisition. Their study, therefore, does not cultivate the mind'.[6] The implication was clearly that they should not usurp in any way the pre-eminence of the classics.

The many pleas for more science were finally heard. The new examination statute of 1849 made the subjects of mechanical and chemical philosophy compulsory for a degree, whilst for honours, in addition to the above, an extended knowledge of one or more natural sciences was obligatory.

A tract written in 1859 – 'Is Educational Reform required in Oxford, and what?' – confirms that before being awarded the BA degree everyone had to pass an examination in natural science. The 'pass man' had to show a general acquaintance with the elements of mechanics, of chemistry and of physiology. The 'class man', besides a higher general knowledge of these three subjects, was required to have an accurate knowledge of some of the branches into which they were divided, e.g. of light, heat, electricity and the three branches of mechanics.

Despite the many petitions for reform of the universities of Oxford and Cambridge that came from within the universities themselves it needed Government authority to break the conservative and traditional chains. The two universities remained aristocratic, closely linked with the Established Church, and associated educationally with classics, theology and theoretical mathematics. Needless to add they were very expensive. Their appeal was mainly to the upper classes; and no dissenters (Non-Conformists) could be admitted. Such a system was opposed by the Liberals and the Non-Conformists, by scientists and by manufacturers.

When the Liberals came to power under Lord John Russell in 1846 they were sympathetic to the appeals for reform and in 1850 appointed a Royal University Commission to look into the two universities. The Commission came out strongly for more research and for extension of the teaching to cover modern developments in science. By the University Act of 1854 for Oxford, and by a similar Act in 1856 for Cambridge, the universities were given the freedom and power to reform themselves – which they did, to become the world's leading centres of learning.

The Oxford Honours School in Natural Science was established in 1852. Another major development in the teaching of physics at the University was the establishment of the **Clarendon Laboratory for Experimental Philosophy**. On 4 February 1868 it was decided[7] that members of the university did not need a riding school, but that the money that had been put aside for this purpose (about £12 000, the profits from the printing of the Earl of Clarendon's manuscripts) be applied instead to the erection of a laboratory for the Professor of Experimental Philosophy – at the time Robert Bellamy Clifton FRS (1836-1921). Clifton, who previously held the Chair of Natural Philosophy at Owens College, Manchester, had accepted the Chair of Physics at Oxford in 1866 and 'at once instituted experimental teaching',[8] beginning with a small class in a room borrowed from the Anatomy Department. Clifton designed and set up what came to be known as the Clarendon Laboratory, in which he taught up to the time of his retirement in 1915. This was the first laboratory in England 'which was specially built and designed for the study of experimental physics'.[9] The laboratory was begun in 1868, first used for classes in 1870, and finished in 1872. Clifton proudly described the layout of his laboratory to the Devonshire Commissioners.

> The building has been divided into three segments, so that it can be used by three or four teachers working simultaneously, with a view to a future extension of the staff. On the north side on the ground floor, are rooms intended for the study of heat, with a balance room attached; on the first floor, on the north side, the rooms are devoted to electricity; and the south side is entirely devoted to optics. There is, besides, the lecture department, represented by a theatre and lecture room and cellars attached.[10]

The building cost between £10 000 and £11 000, not including the apparatus.

Clifton was asked if he had any time from his teaching for original research. He was very definite in his reply.

> Absolutely none. I may mention that during the term in which I have had the laboratory class, I have rarely managed to do the work which I considered necessary in less than eight hours' laboratory work a day, and I was too exhausted at the end of the day to devote myself to research. And even during the terms in which I have merely given lectures, these required so much preparation that I found very little time for doing more than answering the questions of students who called upon me ... . Twelve experimental lectures, if the lectures are tolerably elaborate and cover a good deal of ground, require a large amount of time to be spent in their preparation, especially when the experiments are of a quantitative character.

Much information about the state of physics teaching at Oxford during the 1870s is provided by the Reports of the Royal Commission on Scientific Instruction and the Advancement of Science, known as the Devonshire Commission, established in 1871. The *Third Report of the Commission*, 1873, details the physics instruction at Oxford in 1872 and 1873. The staff involved in teaching physics consisted of the Sedleian Professor of Natural Philosophy (the Rev. B. Price MA FRS),

> who lectures in mechanics and optics 'treated mathematically'; the Professor of Experimental Philosophy (R.B. Clifton, M.A., F.R.S.), 'who teaches physics mainly experimentally, i.e. lectures are illustrated by experiments and instruction in experimental methods is given in the laboratory'; the Waynflete Professor of Chemistry, a recent Chair established in 1854; and the Savilian Professor of Astronomy (C. Pritchard, M.A.), together with Demonstrators in Experimental Philosophy and in Chemistry.

Clifton described his courses of instruction in physics to the Devonshire Commissions (July 1870) as follows:

> I have really two kinds of instruction. When the students first came they generally knew little about the subject, but they probably know some

> mathematics, some knowledge of which is essential. Then they attend, first of all, a course of experimental lectures, by which they become accustomed to seeing experiments performed; and also get some general knowledge of the subject; I also recommend some textbook which they are to read. If they intend to continue in physics for their special honour subject, they come to the physical laboratory to study there in quite a different way. In the laboratory they have apparatus themselves, each student works separately, and they are called upon to perform certain experimental processes, mostly quantitative determinations, the teacher of course giving them full instructions in the detailed use of the instruments. Afterwards, researches or portions of researches, which have been performed, are repeated. And with the increased accommodation which is now to be provided it will eventually enable the advanced students to do original work.[11]

On average about 40 students attended Clifton's experimental classes.

During the Michaelmas Term 1872, Price, the Sedleian Professor of Natural Philosophy, was giving lectures in hydro-mechanics at the University Museum three times a week. He lectured on the dynamics of rigid systems at the same location and same times during the Hilary Term 1873, while Pritchard, the Savilian Professor of Astronomy gave a similar number of lectures on the Ninth and Eleventh Sections of Newton's *Principia* and on the lunar and planetary theories, at the University Museum. During the Trinity term following Price lectured on ... problems in applied mathematics. His class averaged 15–20 students. In addition, as the Commissioners point out, 'a very large amount of private instruction, in addition to the lectures, is, of course, given by the lecturers in their respective colleges'.[12]

Christ Church could boast three Readers in science – one each in physics, in chemistry and in physiology. During the academic year 1872–3 A.W. Reinold, MA, the Reader in Physics, gave courses in heat and electricity in the Christ Church Laboratory on Mondays, Wednesdays and Fridays at 10 a.m.

To economize on scarce resources, both human and material, in the matter of practical work in the sciences, the Commissioners argued that:

> the most convenient arrangement would probably be that the more elementary instruction should be given in the college laboratories, and that when the student had attained a certain degree of proficiency he should pass under the hands of the Professors in the central laboratories of the University.

Even though it might not be possible for each college to have its own separate laboratories, the Commissioners were in no doubt that each college should have its own lecturers in Natural Science.

To prevent such a situation as we have today in this age of narrow and early specialization, especially in the sciences, the Commissioners argued that a wide spectrum of knowledge, covering both the sciences and the humanities, should be the hallmark of anyone professing to have spent three or more years at a university. The Commissioners argued that whilst the science student should have some acquaintance with literature, the student of Classical Literature or of Theology should also show some evidence of scientific culture. Professor Price had informed the Commissioners that it was his view that 'the elements of science ought to be known by all students at Oxford'.[13]

Cambridge

The tradition at Cambridge of emphasis on the mathematical and physical sciences that stemmed from Newton was maintained throughout the 19th century. Reporting in 1835 the *Quarterly Journal of Education* asserted that 'the Cambridge course of education includes a tolerably complete course of natural philosophy, perhaps the most complete of any mathematical school in the world.'[14]

The Cambridge Calendars of the first half of the 19th century show clearly that the Jacksonian Professors of Natural and Experimental Philosophy and the Plumian Professors of Astronomy and Experimental Philosophy gave experimental lectures on mechanics, hydrostatics, optics, astronomy, magnetism and electricity. Francis Wollaston, the Jacksonian Professor at the start of the century exhibited 'not less than three hundred experiments annually'. Wollaston was succeeded as Jacksonian Professor in 1813 by William Farish who held the Chair till 1837. Farish was then succeeded by the Rev. R. Willis. In his evidence to the Devonshire Commission Willis said this of his predecessor.

> Professor Farish lived just at that era when machinery was being developed for the purpose of manufactures, and he contrived a system of apparatus by which he could build up the machines that were required from day to day for his lectures, and he did that with great ingenuity and success.

That there was considerable scientific activity in Cambridge is evidenced by the fundamental researches and experiments being undertaken by several members of the university. While still an undergraduate at Emmanuel College, Thomas Young (who gave us the wave theory of light) wrote 'On Sound and Light' for the Royal Society. William Hyde Wollaston of Caius College, brother of the Jacksonian Professor, was the first to use a total reflection method of obtaining refractive indices. In 1820 the Rev. William Cecil, of Magdalene College, gave an account of his invention of an Internal Combustion Engine to the Cambridge Philosophical Society which was founded in 1819.

A feature of the period was the emergence of the private tutor or coach. These tutors did much to further knowledge of physics not only by their instruction but also by the many textbooks they wrote. The large number of physics textbooks that came to be written by members of the university to cater for the demands that had grown up for them is also indicative of the scientific activity that prevailed in Cambridge. In 1837 William Whewell's *Mechanical Euclid* appeared. Whewell MA, FRS (1794–1866), was Fellow and Tutor of Trinity College. This book dealt with those topics of mechanics and hydrostatics that were required for the BA degree, as follows:

> *Statics:* Definitions and fundamental notions and axioms. The lever, the composition and resolution of forces, the wheel and axle, pulleys, the inclined plane, virtual velocities, centre of gravity, pressures on fulcra.

Hydrostatics: Definitions, fundamental notions, axioms. Pressure of fluids, specific gravity, the hydrometer, weight of the air and the barometer, the siphon, the common pump, the forcing pump, the air-pump, the condenser, the thermometer.

The Laws of Motion and their consequences.

Clearly the standard in statics and hydrostatics for the general degree was no higher than that required by the present-day GCSE student. However, it must be remembered that the age of undergraduates at Oxford and Cambridge was often no higher than that of today's GCSE students. If anything it was probably lower. Dr J.A. Voigh, writing of his tour of Britain in 1855, remarked that boys entered the university at 15, and stated that the average age in 1831 was as low as 13. It must also be remembered that this requirement was the minimum required in physics from *all* degree students, not only from those specializing in the subject. *Every* student had to have some acquaintance with physics, which is a far more desirable state of affairs than applies in these days of narrow specialization where acquaintance with even elementary physics is not obligatory on all degree students.

In 1836 John Henry Pratt, Fellow of Gonville and Caius, published his *Mathematical Principles of Mechanical Philosophy* for the use of university students. The first 1000 copies were 'sold off sooner than was expected ... and the demand for copies being pressing' a second edition was issued in 1842. This was an advanced treatise, employing the calculus to solve problems of statics, dynamics, hydrostatics and hydrodynamics.

In 1828 Sir George Airy was appointed to the Plumian Chair of Astronomy and Experimental Philosophy. His lectures, which were spoken of highly, embraced mechanics, hydrostatics, pneumatics and optics. They included up-to-date information; for instance, the undulatory theory of light was demonstrated by means of experiment in the course of his lectures. Whewell remarks that Airy's lectures were attended with great avidity.[15]

On his retirement in 1836, Airy was succeeded by James Challis. Challis is described as giving 20 lectures a year, although his class seemingly only averaged about five students.

As at Oxford, petitions for reform of the university regulations at Cambridge came from within the university. In 1842 a new tripos was established in natural science. In 1845 Whewell, who was now Master of Trinity College, argued in his essay 'Of a Liberal Education', that a study of modern experimental and inductive science, embracing subjects such as mechanics, hydrostatics, astronomy, optics, would fittingly supplement the knowledge of ancient literature as part of a liberal education. To effect this he outlined a plan of a standard Cambridge Course of Mathematics. This plan is given below, since it was adopted by Whewell's college and eventually by the university as a whole.

1. Arithmetic
2. Algebra
3. Trigonometry
4. Conic Sections
5. Mechanics: For this subject he recommended his own *Elementary Treatise on Mechanics.*

   Statics: The lever; composition and resolution of forces; equilibrium of a rigid body; analytical formulae of statics; centre of gravity; simple machines.

Dynamics: Accelerating force; laws of motion; uniformly accelerated motion; motion by gravity; variable forces; projectiles; centrifugal force; general formulae; moving force; motion on inclined planes; motion on a curve.

6. Newton's *Principia*
7–9. Calculus
10. Hydrostatics: Webster's *Principles of Hydrostatics*
11. Optics: Dr Wood's *Elements of Optics*
12. Astronomy: Dr Hymer's *Astronomy*

Whilst adequate instruction was being given in mechanics, hydrostatics, optics and astronomy, the other branches of physics, such as heat, magnetism and electricity were not neglected. It was to foster still further the study of these subjects that the Examinations Syndicate recommended in 1848 that all candidates for the BA degree, in addition to what was already required in mechanics and hydrostatics, must have attended satisfactorily, for one term at least, the lectures given by the Professor of Chemistry or by the Jacksonian Professor of Natural and Experimental Philosophy. These lectures included the subjects of heat, magnetism and electricity. This proposal was approved as a condition for the general degree, but it was decided not to add the subjects of heat, electricity and magnetism to the Honours Examination in view of 'the great number of subjects [statics, dynamics, hydrostatics and astronomy] which now occupy the attention of the candidates'.

The Jacksonian Chair of Natural and Experimental Philosophy and Chemistry (founded in 1783) was held at the time by the Rev. R. Willis. The Report of the Commissioners that inquired into the university in 1852 paid a glowing tribute to the work of Willis in spreading knowledge of physical science. The Report states that Willis,

> by selecting the science of Mechanics generally, more especially viewed in connection with machinery and engineering, as the principal subject of his Lectures, accompanied by very elaborate models and experimental illustrations, has carried out, with remarkable ability and success, the primary design of this foundation [Jacksonian Professorship].[16]

The Commissioners also singled out the Lucasian Professor of Mathematics – G.G. Stokes (appointed in 1849) for praise. Stokes gave lectures, illustrated by means of experiments, on 'some of the more important and difficult propositions in Optics, Mechanics, Hydrostatics and Hydrodynamics'.

In his evidence to the Commissioners Stokes stated that opportunity was afforded to the students themselves 'in some cases of performing a variety of experiments'.[17] However, Lord Rayleigh (1842–1919), who attended the lectures of Stokes on optics and was clearly impressed by the lecture demonstrations, confessed he had no experimental training while at Cambridge. As he says, he simply dabbled in chemistry and electricity on his own account. He stated later that this lack of instruction in laboratory work wasted three or four years of his scientific life.

That laboratory work by the pupils themselves was not a regular part of the instruction in physics was not due to the pedagogues being unaware that the experimental method was fundamental to the teaching of science, but to the difficulty of procuring apparatus. Air-pumps, telescopes, even thermometers, were expensive luxuries.

James Clerk Maxwell

From the early education of James Clerk Maxwell we glean much information as to the scientific education available to young men at that time. Maxwell (1831–79) was born in Edinburgh. He was sent to Edinburgh Academy which had been founded some years earlier by Sir Walter Scott among others. When he was 11 years old he had a great treat of being taken 'to see electromagnetic machines'. By age 14 he was attending the meetings of the Royal Society of Edinburgh. When he was barely 15 he communicated his first published paper to the Society: 'On the Description of Oval Curves and these having a Plurality of Foci. With Remarks by Professor Forbes'. At the time of this paper Maxwell had received no instruction in mathematics beyond a few books of Euclid and the merest elements of algebra.

Also at school at the Academy was P.G. Tait, later to become Professor of Natural Philosophy at the University of Edinburgh. Clerk Maxwell and Tait were acknowledged as the two best mathematicians in the school. Believing that the pupils should have some acquaintance with physical science, one of the classical masters made good this deficiency by giving lessons 'out of a book'. Campbell writes, 'The only thing I distinctly remember about these hours is that Maxwell and Tait seemed to know much more about the subject than our teacher did.'[18] Maxwell was given a prism by Nicol. He was soon working at magnetism and the polarization of light.

At the age of 16 Maxwell entered the University of Edinburgh, learning mathematics under Professor Kelland, natural philosophy under Professor Forbes and logic under Sir W.R. Hamilton. Maxwell described his studies.

> In Forbes we do the lever, which is all in Potter's *Mechanics* ... . On Monday, Wednesday, Friday, I read Newton's Fluxions in a sort of way... . At 11 there is Forbes, who has now finished the introduction and properties of bodies, and is beginning Mechanics in earnest ... . Then I extend notes, and read textbooks, which are Kelland's *Algebra* and Potter's *Mechanics* ....[19]

It was Forbes who developed in Maxwell a practical down-to-earth interest in experimental techniques rare in a theoretical physicist. In a letter to his friend Lewis Campbell, dated July 1848, Maxwell describes a makeshift laboratory he had improvised in a garret.

> I have regularly set up shop now above the wash-house at the gate, in a garret. I have an old door set on two barrels and two chairs, of which one is safe, and a skylight above which will slide up and down. On the door (or table) there is a lot of bowls, jugs, plate, jam pigs, etc. containing various chemicals; also broken glass, iron, and copper wire ... sealing wax ... a lens, a Smee's galvanic apparatus, ... and a countless variety of little beetles etc. ... which fall into the different liquids and poison themselves ... . I intend to set up some more galvanism in jam pigs ... so I am experimenting on the best methods of electro-typing ... I melt out the wax with the lens that being the cleanest way of getting a strong heat ... .
>
> I took down the clay gun and made a centrifugal pump of it; also tried experiments on sound under water, which is very distinct, and I can understand how fishes can be stunned by knocking a stone ... Electric Telegraph. This is going so as to make a compass spin very much.

I can now polarise light by reflection or refraction in four ways, and get beautiful but evanescent figures in plate glass by heating its edge.[20]

In his third year, in addition to attending Forbes' lectures, Maxwell was also attending classes in chemistry (Professor Gregory), in practical chemistry (Mr Yemp) and in moral philosophy (Professor Wilson). While in his third year Maxwell's paper on 'Rolling Curves' was read before the Edinburgh Royal Society 'for it was not thought proper for a boy in a round jacket to mount the rostrum there'.[21]

In 1850 Maxwell went to Cambridge and began the rigours of preparing for the mathematical tripos, then at the height of its competitive glory. He was 19 and one year older than most of the entering undergraduates. P.G. Tait, who was a year behind Maxwell at school, had gone up to Cambridge two years earlier. He was Senior Wrangler in 1852. Tait, who was also with Maxwell in the classes of Forbes and Kelland at the University of Edinburgh, describes how Maxwell spent three years at the University, working (without any assistance or supervision) with physical and chemical apparatus, and devouring all sorts of scientific works in the library. Tait continues:

> Thus he brought to Cambridge, in the autumn of 1850, a mass of knowledge which was really immense for so young a man, but in a state of disorder appalling to his methodical private tutor. Though the tutor was William Hopkins, the pupil to a great extent took his own way, and it may safely be said that no high wrangler of recent years ever entered the Senate House more imperfectly trained to produce 'paying' work than did Clerk Maxwell. But by sheer strength of intellect, though with the very minimum of knowledge how to use it to advantage under the conditions of the examination, he obtained the position of Second Wrangler, and was bracketed equal with the Second Wrangler in the higher ordeal of the Smith's Prizes.[22]

During his first year at Cambridge Maxwell attended the Trinity College lectures on Mathematics. As they were somewhat elementary he worked harder problems as a private pupil of Mr Porter of Peterhouse, his private tutor. It was in his second year that he became a pupil of Hopkins, the great coach. In a diary kept by a Mr Lawson, he records:

> He [Hopkins] was talking to me this evening about Maxwell. He says it appears impossible for Maxwell to think incorrectly on physical subjects ... He looks upon him as a great genius with all its eccentricities, and prophesies that one day he will shine as a light in physical science.[23]

Maxwell also attended the lectures of Professor Stokes, the Lucasian Professor of Mathematics.

Maxwell graduated as Second Wrangler in 1854. In the same year he undertook the job of external examiner at Cheltenham College. Maxwell was at Cambridge for the May term 1855, preparing for the Fellowship examination for Trinity College. He read Poisson's papers on electricity and magnetism and began putting together more systematically his own ideas on Faraday's Lines of Force. He was also working on the motion of liquids and at his colour-top. He was duly awarded a Fellowship at

Trinity and was at once appointed to lecture to the upper division of the third year in hydrostatics and optics. in September 1855 he wrote to his father: 'I am getting my electrical mathematics into shape and I see through some parts which were rather hazy before; but I do not find very much time for it at present, because I am reading about heat and fluids, so as not to tell lies in my lectures.'[24] In November he wrote: 'I have been lecturing two weeks now, and the class seems improving; and they come and ask questions, which is a good sign. I have been making curves to show the relations of pressure and volume in gases, and they make the subject easier.'

Maxwell also had a large share in 'exercising the questionists', i.e. preparing the pass men for the final examination by setting papers in arithmetic, algebra etc. and going over them with the students individually.

At this time, the young Clerk Maxwell thought it worthwhile to attend the lectures of the Professor of Mechanics (Willis) and to exchange ideas with him. In a letter to his father in November 1855 Maxwell wrote:

> I attended Willis on Mechanism today, and I think I will attend his course, which is about the parts of machinery. I was lecturing about the velocity of water escaping from a hole this morning. There was a great noise outside and we looked out at a magnificent jet from a pipe which had just gone wrong in the court. So that I was saved the trouble of making experiments ...
>
> I have got a contrivance made for expounding instruments. It is a square rod, one yard long, on which slide pieces, which will carry lenses. Each piece has a wedge which fixes it tight on the rod, and a saw-shaft, with holes through it, for fastening the plasterboard frame of the lens. By means of this I intend to set up all kinds of models of instruments.

The Rev. Charles Hope Robertson, who was at Trinity in the same years as Clerk Maxwell, wrote thus of Maxwell.

> As an original experimenter he was most ingenious in contriving out of simple means apparatus for delicate experiments. I have to this day crystals for showing polarised light, which he gave me, cut and polished by most simple rubbing, mounted with cardboard and sealing wax.
>
> His essay on the Rings of Saturn – showing how mechanical principles required that these bodies were not solid, but formed by multitudes of small bodies revolving round Saturn, in bands of orbits – has received abundant confirmation from recent observation with large telescopes. His own simple experiments with corks and rings suggested the idea.
>
> But while so ingenious himself he had great difficulty in imparting his ideas to others; consequently was not so clear a lecturer or writer as might have been expected. It was probably this that prevented his being Senior Wrangler of his year.[25]

On 13 February 1856 Professor J.D. Forbes of the University of Edinburgh wrote to Maxwell suggesting that he should apply for the Chair of Natural Philosophy at Marischal College, Aberdeen. He was appointed to the college and remained there until 1860, when he was appointed Professor at King's College, London. However,

Maxwell retained his connection with Cambridge as a Moderator in the mathematical tripos during 1866–70. In 1871 Maxwell returned to Cambridge as the first Professor of Experimental Physics and began planning the Cavendish Laboratory.

The teaching of mathematics at Cambridge at this time was described by R.T. Glazebrook as 'one-sided'. He noted the complaint that Cambridge was behind the rest of the scientific world, and that, whereas the students of so many other universities were introduced to the splendid discoveries of such subjects as electricity and heat, the wranglers of Cambridge spent their time upon mathematical trifles and problems, so called, barren alike of practical results and scientific interest.

Things changed for the better with the appointment of Sir William Thomson, Professor of Natural Philosophy at Glasgow and later Lord Kelvin, as Public Examiner and Maxwell as Examiner and Moderator. As Glazebrook says: 'Maxwell's questions [as Moderator in 1866] infused fresh life into the Cambridge Tripos, and, therefore, into the University studies, by the number of original ideas and new lines of thought opened up by them.' In 1868 the examination for honours in the mathematical tripos was extended to include the subjects of heat, electricity and magnetism. Maxwell was asked one day whether a certain physical law was true or not. He replied: 'It is true in the three days, but not in the five', alluding to the then existing division of the mathematical tripos into three days of elementary papers and five days in which a more advanced knowledge was expected of the candidate.

As in the case of Oxford University, the Report of the Royal Commission on Scientific Instruction and the Advancement of Science tells us much about the state of physics teaching at the University of Cambridge during the 1870s. The staff engaged in teaching various aspects of physics at Cambridge at the time of this Royal Commission were: J. Challis, MA, the Plumian Professor of Astronomy and Experimental Philosophy; R. Willis, MA, the Jacksonian Professor of Natural and Experimental Philosophy; J. Clerk Maxwell, the Professor of Experimental Physics; G.G. Stokes, MA, the Lucasian Professor of Mathematics; and G.D. Liveing, MA, the Professor of Chemistry, assisted by a demonstrator.

University lectures were given in the Astronomical Lecture Room, the New Museums and the Chemical Lecture Room, and courses in practical chemistry in the University Laboratory. The lecture titles during the three university terms included: Practical Astronomy and Magnetism, Mechanics and the Principles of Mechanism, the Steam Engine etc., Heat and Electricity, Electrostatics, Electromagnetism, and Hydrostatics, Hydrodynamics and Optics.

In addition to these university courses, physics instruction was given by certain colleges which had combined into groups. When asked by the commissioners if he had a laboratory and apparatus with which to illustrate his lectures on heat and electricity, Mr Trotter, one of the college lecturers, replied:

> No; most of the apparatus that I have got is my own. The college [Trinity] has fitted up a lecture room for me, with gas and water, etc, and has given a grant for apparatus, and I have also got a good deal of apparatus of my own. I am not able to illustrate my lectures as fully as I should like, but still I am able to show a good many illustrations of the points which are theoretically most important.

The Rev. H. Latham, MA, another college lecturer, stated in his evidence:

> In my own lectures, in physical subjects, even for the ordinary degree, I endeavour to show, experimentally, everything that I possibly can. I lecture on hydrostatics and optics, and I try to exhibit all the phenomena which I wish my class to calculate and to explain. It is very advantageous, even for a pass student, to give him a hydrostatic balance and set him to work to find the specific gravity of a substance. They will often do this with considerable accuracy, and understand very much better than if they had merely learnt from a book.

In his evidence to the Devonshire Commissioners (29 November 1870) the Rev. J. Challis, MA, FRS, pointed out the great need at Cambridge for the hardware to make the experimental teaching of physics at the university more of a reality. 'I think that the first and foremost want at Cambridge has been the means of making experiments, the apparatus for experiments, and buildings for experiments.'[26]

Admittedly the University had erected a Museum Building about eight or nine years earlier for the teaching of science but Challis pointed out that the building, crowded with mineralogical, anatomical and botanical collections, was 'not adequate for the purpose of teaching physics ... there are no conveniences for making experiments in physics. We want a new building for that purpose'. That new building was the Cavendish Laboratory.

J.J. Thompson, who later was to become one of the great Cavendish Professors of Experimental Philosophy, came to Cambridge in 1876 to prepare for the mathematical tripos, which then covered a very wide range of pure and applied mathematics, leaving him no time for experimental work. He stated that the physical subjects were considered much too exclusively from the purely mathematical side. His writings give us an illuminating account of the teaching of physics at Cambridge at the time.

> I left Manchester and came into residence in Cambridge in 1876 ... The lectures of the mathematical professors in those days were not attended by many undergraduates; at one course of lectures given by Professor Cayley I was the only undergraduate present.
>
> I also attended the lectures of Professor Adams. His lectures were wonderfully clear and were read from beautifully written manuscripts, which he brought into the lecture-room in small calico bags.
>
> Of all the lectures I attended I think I enjoyed Sir George Stokes's the most; they were much more physical than the others and were illustrated by experiments which succeeded with a precision which I have often envied since when I have had to do lecture experiments myself. Stokes had, I think, an unrivalled power of explaining the essentials of a physical problem in a clear and simple way, very often without the use of mathematical symbols. In fact, he practised Bacon's aphorism that the physics ought to precede the mathematics, and often had developed the main features of the problem from physical principles before he began the analytical investigation. His interest in his subject sometimes made him forget that the canonical time for a lecture is one hour; but I do not remember that he ever went on so long as he did on one occasion many years after, when he was over eighty years of age. One afternoon at half-past two

I met a batch of my students coming into the Laboratory and looking very hungry; I asked them what they had been doing, and they said they had just come from Sir George Stokes's lecture, which had commenced at twelve.

I was much interested in, and derived great benefit from, the lectures on mathematical physics given in Trinity College by Sir W.D. Niven, and those on pure mathematics by Dr Glaisher. Some of the College lectures were, I must acknowledge, the reverse of interesting; one lecturer adopted a method I have never seen before or since. He hardly spoke a word, but wrote steadily on the blackboard; when he had filled it he told us to copy that down, and waited in silence until those who obeyed him had done so; then he cleaned the board and began again. Like nearly everyone else, I 'coached' with Dr Routh during the time I was preparing for the Tripos. Routh's teaching was not the least like what is ordinarily understood by coaching; it was really an admirably arranged series of exceedingly clear lectures given to a much larger audience than most of the lectures of the University professors or College lecturers.

My connexion with the Cavendish Laboratory began in 1880, immediately after taking my degree. Lord Rayleigh had just been elected to the professorship. ... I turned my attention, at Lord Rayleigh's suggestion to the determination of the ratio of the electrostatic to the electromagnetic unit of electrical charge. Other subjects at which I worked were the Electromagnetic Theory of Light, the Properties of Moving Electrified Charges, Vortex Motion, and the theory of Electric Discharge through Gases.

There was great activity at the Laboratory during Lord Rayleigh's tenure of the professorship. Courses for teaching practical physics were organized by Glazebrook and Shaw, and a scheme of lectures in physics for students preparing for the Natural Sciences Tripos was arranged. It was at this time that the classical determinations by Lord Rayleigh of the absolute values of the fundamental units of electricity were carried out.[27]

University of London

University College opened at Gower Street, in October 1828, with over 300 students, nearly 80 having enrolled for the natural philosophy class. The college was run as a government stock company, with shareholders. Being an undenominational teaching institution, with low fees and a wide curriculum, the college was an immediate success. Despite jibes about the Cockney College and of 'Stincomalee' (it was built on a rubbish dump), within a year it had doubled its numbers. In 1831 the College received its grant of incorporation which enabled it to award degrees, despite the opposition from Oxford and Cambridge. In a few decades University College had as many science students as Oxford and Cambridge combined.

No charge of neglecting the sciences could be brought against this University. On the contrary, the *Monthly Review* for January 1829 complained that 'Looking at the list of lecturers and lectures, it is impossible not to be struck with the preponderance which is given to natural and scientific philosophy over moral, which it seems has almost escaped the attention of the presiding powers'.[28]

The first Professor of Natural Philosophy was the Rev. Dionysius Lardner (1793–1859), a man of unbounded energy, who attracted large audiences to his lectures. In his Inaugural Address on the Advantages of Natural Philosophy and Astronomy he asserted that neither attention nor cost had been spared to provide the Natural Philosophy class with apparatus, 'not only in number and variety, but also in magnitude and excellence, equal, if not superior, to any collection in these kingdoms'. This apparatus consisted of air-pumps, balances, electrical and galvanic machines, working models of machines, sectional models, etc. Lardner went on to state that his course would cover mechanics, properties of matter, hydrostatics, hydraulics, pneumatics, astronomy, geology, optics, heat, electricity and magnetism. Furthermore, 'the experiments which indicate the mutual relation of electricity and galvanism to magnetism, will be subjects of careful examination, and with a view to this, a most effective system of apparatus has been prepared.'

Lardner resigned from the Chair of Natural Philosophy in 1831. But he continued to further the cause of physics teaching by the several excellent textbooks he wrote. His extremely good and well-illustrated *Handbook of Natural Philosophy* was published in 1856. It consisted of four treatises; one on mechanics, one on hydrostatics, pneumatics and heat, another on optics and finally one on electricity, magnetism and acoustics. In 1857 appeared Lardner's *Natural Philosophy for Schools*. This excellent textbook for 'junior students' covered the entire range of physics and gave details as to how the experiments were to be performed. By 1874 14 000 copies had been sold.

Lardner was succeeded by Dr William Ritchie (1790–1837), formerly Professor of Natural Philosophy at the Royal Institution. Ritchie was noted for his skill in devising and demonstrating experiments. Lord Brougham asserted that Ritchie was excelled by no living teacher in his ardour to impart knowledge. It was during Ritchie's professorship that in 1839 the university started evening courses in physics for the benefit of masters and ushers of schools. More than 40 teachers enrolled for the first course.

From 1840 to 1865 the Chair of Natural Philosophy was held by Richard Potter, who was the author of several textbooks on *Optics*, on *Hydrostatics* and on *Mechanics*. It appears that Potter was not a particularly competent teacher, particularly towards the end of his tenure.

The subjects of heat, light, electricity and magnetism were also lectured upon by the Professor of Chemistry. The first Professor of Chemistry was Edward Turner, who was described as 'a lucid lecturer' by one of his pupils. In 1837 Turner was succeeded by the celebrated Thomas Graham (1805–69) who held the Chair till 1855.

In 1840 a Chair of Civil Engineering was established. The Course for civil engineers embraced mechanics, dynamics, hydrostatics, hydraulics, pneumatics, the steam engine, optical instruments and heat.

The *Fifth Report of the Royal Commission on Scientific Instruction and the Advancement of Science*, 1874, Vol. XXII provides useful information about University College, London. Physics teaching was given by the Professor of Physics, by the Professor of Applied Mathematics and Mechanics and by the Professor of Chemistry and Applied Chemistry. The Commissioners reported that

> the courses of study, as indicated in the programmes of the professorial lectures appear to be carefully arranged with a view to the requirements both of elementary and of advanced students. Great importance is attached to the laboratory instruction in Physics, Chemistry and Physiology.[29]

The Professor of Physics certainly enjoyed no sinecure. He gave six lectures a week, that is, 200 lectures per session. He conducted four Exercise Classes a week, or 130 per session. He was occupied in the Laboratory daily from 10 till 5. 'The Professor is more or less occupied daily in superintending the students.'[30] And for his services he was 'very inadequately paid'. The Report points out the consequences of inadequate remuneration – 'in many cases the College has found it impossible to retain the services of some of its most distinguished Professors.'[31]

Professor G. Carey Foster, BA, FRS, Professor of Physics at University College, was adamant that all serious students of physics should pass through a practical course.

> I was convinced that there could be no sound teaching of physics apart from practical work; students must have personal acquaintance with phenomena before they can profitably reason about them ... .
>
> Instruction in physics is given in several classes – in addition there is instruction in a physical laboratory. The instruction is chiefly based on experimental demonstrations.
>
> The physical laboratory has only existed for the last five years. The instruction in physics formerly given was confined to lectures. A large lecture room was provided, and a room for keeping apparatus, but no place in which it was intended that experimental investigations should be made, or practical instruction should be given ... . The subjects in which most of the students have worked in the laboratory are heat, magnetism, and electricity. Frequently one of the first things that I put them to, in order to use what they already know, is to determine a few specific gravities, and some of the simplest physical measurements of that kind; but heat, electricity, and magnetism are the staple subjects of the work.[32]

He made a special plea for additional assistance since 'there is work enough for perhaps half a dozen besides myself'.

King's College

King's College, London, was established in 1831 by the Tory Party (mainly Church of England), to counterbalance University College, the 'Godless institution' of the Non-conformist and Whig parties. Among the first Senate members were Faraday and John Dalton.

The first Professor of Natural and Experimental Philosophy (1831) was the Rev. Henry Moseley FRS (1801–72). He established the basic courses in properties of matter, statics, dynamics, hydrostatics and hydrodynamics, pneumatics and the application of mechanical science in arts and manufactures. In addition, Moseley gave the evening course on the theory of heat and the steam engine.

Moseley resigned in 1844 to take up an appointment as Inspector of Schools (see Chapter 6). He was succeeded by the Rev. Matthew O'Brien.

In 1834 a special Chair of Experimental Physics was created, to which Sir Charles Wheatstone (1802–75) was appointed. A brilliant researcher who attracted many students to the College, Wheatstone had an almost pathological fear of addressing an audience, and as a result his lectures did not match his brilliance in experimental work. After 1840 most of his teaching work was done by other lecturers.

The Professors of Chemistry at King's College during this period were John Frederic Daniell (1790–1845) and his successor W.A. Miller. Both men included physics subjects such as mechanics, properties of matter, light and electricity and heat in their lectures.

The high reputation for science enjoyed by King's College in the 1860s owed something to the fact that James Clerk Maxwell took up the Chair of Natural and Experimental Philosophy in 1860. Maxwell was a brilliant physicist but he was not a good teacher of raw youths. While he was in London Maxwell resided at 8 Palace Gardens Terrace, Kensington. It was here that he carried on many of his experiments in a large garret which ran the whole length of the house. When experimenting at the window with the colour box (which was painted black and nearly eight feet long), he excited the wonder of his neighbours, who thought him mad to spend so many hours in staring into a coffin.[33]

In 1865 he left the College for health reasons and retired to his country estate in Scotland to carry on his researches into electromagnetism. He was succeeded by W.G. Adams.

King's College, London, has the distinction of being 'the earliest institution in which laboratory physics was pursued according to a systematic plan for its educational value, and was a required part of the work necessary for a degree.'[34] Professor Clifton, the first professor of Natural Philosophy at Owens College, Manchester, had suggested specially designed laboratories for the study of experimental physics. This suggestion was adopted by King's College in 1868.

In his evidence to the Devonshire Commission Professor Adams describes this laboratory 'for the teaching and practical study of physics'.

> Adjoining the museum, which is 66 feet by 30 feet, with cases all round the walls full of physical apparatus, two other rooms were set apart, one is 30 feet by 18, and another 30 feet by 40, the first for a laboratory for the Professor to prepare his experiments, and for the more delicate experiments made by the students, and the other for the general experiments to be made by the students. In addition to these, another room was built for a battery room and storeroom, and a lecture room capable of containing 80 students was set apart for experimental physics.[35]

Adams goes on to describe the kind of practical work done by the students.

> The physical laboratory is adapted to the wants of those students who wish to acquire a thorough practical knowledge of pneumatics, heat, light, electricity, and magnetism. The course of practical work for any student depends on his previous knowledge; and he may pursue any one subject in detail, so as to obtain a thorough knowledge of its principles and practices. Students are taught to use the barometer, thermometer, cathetometer, and hygrometers; to determine the

specific gravities of solids and liquids; and their specific and latent heats; to measure melting and boiling points; to measure relative powers of conduction of bodies for heat and for electricity; to measure the focal lengths of mirrors and lenses, the indices of refraction of various transparent substances, their powers of dispersion, and the relative distances of the fixed lines of the solar spectrum. They study the principles of the polarization of light, and the magnetic and electrical properties of bodies, the different methods of developing electricity; and learn to determine the electromotive force and resistance of batteries, the strength of currents, the resistance of telegraph-wires in absolute measurement; and are taught the testing of cables, and the principles of telegraphy ... . In addition the professor and lecturers also give instruction in evening classes.

These evening classes at King's College were eminently successful, providing a complete course of science instruction for persons unable to attend the day classes. In 1873 the evening classes were attended by as many as 550 students, about 300 of whom attended the science classes.

In 1836 the two separate university colleges were amalgamated to form a federal University of London. The first examinations for degrees in the new University of London were held in 1839. Candidates for the BA degree were required to show a competent knowledge in the four branches of the examination, namely, mathematics and natural philosophy, animal physiology, classics, and logic and moral philosophy. The natural philosophy embraced mechanics, hydrostatics, hydraulics, pneumatics, heat, optics, astronomy and electricity.

In 1857 memorials were presented to the University of London to institute degrees in science. In 1860 the Faculty of Science of the University of London was established. From this time on the degrees of BSc and of DSc were awarded for the first time in England.

The aim of the BSc degree was to 'lay a broad foundation of Scientific Culture'. The First (Pass) examination for the degree of BSc included pure mathematics, experimental physics, inorganic chemistry and biology. The Second (Pass) examination included mechanical and natural philosophy, organic chemistry, geology, physiology, logic and moral philosophy. Later on, honours could be taken in any subject. The degree of BSc of London University carried great prestige in the educational world, for many years having no serious rival.

At the time there were no practical examinations in physics for the Intermediate and Final BSc examinations nor for the preliminary examinations for medical degrees. This shortcoming came in for much criticism. It was not till 1889 that a '*viva voce* interrogation on the use of instruments' was made part of the Intermediate BSc examination in physics and that a practical examination in physics was compulsory for all Final BSc candidates.

A medical department formed part of London University, and since a knowledge of physics was considered indispensable to a future physician, numerous medical students received instruction in the subject as part of their professional education. In addition to University College and King's College, and in addition to the medical schools of the large London hospitals, medical schools existed in almost all large provincial cities. These schools were affiliated to the University of London and took its examinations.

The Provincial Universities

Owens College, Manchester, was started in 1851 as an affiliated College of London University. Its students took the science degrees of London. The first occupant of the Chair of Natural Philosophy was Robert Bellamy Clifton (1836–1921). He held the Chair till 1866 when he was tempted to the Chair of Physics at Oxford. Balfour Stewart was appointed Professor of Natural Philosophy in 1870. Among his pupils were J.J. Thomson, Poynting and Arthur Schuster. J.J. Thomson gives this account of the physics teaching at Owens College under Professor Balfour Stewart.

> Fortunately before coming to Cambridge I had had at the Owens College, Manchester, an opportunity of working in a physical laboratory under Professor Balfour Stewart. When I was at the Owens College in the middle seventies physical laboratories were not so crowded as they are nowadays, and the students were allowed to work more independently than is possible where the numbers are large. We were allowed considerable latitude in the choice of the experiments we made, we set up our own apparatus and spent as much time as we pleased in following any points of interest that turned up in the course of our work; the work was thus from the first more interesting and more educational than under the highly organized systems of instruction which a large number of students necessitates.
>
> Professor Balfour Stewart, in addition to being a most excellent teacher, was enthusiastic about research and succeeded in imparting the same spirit to his pupils. I remember that one day, when I had only been a short time in the Laboratory, he was talking about sun-spots and he said that he had a large number of observations which he thought might throw some light on the connexion between sun-spots and terrestrial phenomena, but that he had not the time to reduce them. I ventured to say that if I could be of any help I should be glad to do what I could, and he gave me a number of observations to reduce. Though the work I did was purely arithmetical, I found it very interesting, and enjoyed the feeling that I was taking some part in real science, a feeling which is not given by the repetition of a laboratory experiment.
>
> Balfour Stewart was much interested in the question as to whether there is any change in weight when substances combine chemically. When I was helping him with some experiments on the combination of mercury and iodine these substances combined with such violence that the flask they were in exploded in my face and I nearly lost the use of both eyes.
>
> Whilst I was in the laboratory at Manchester, Schuster was working at the spectroscope; he gave a course of lectures on Maxwell's *Electricity and Magnetism* which I attended, and Poynting, another old student, returned to it for a short time after taking his degree at Cambridge.[36]

Balfour Stewart's work on radiation and spectrum analysis was to inspire the researches of Crookes, Liveing, Dewar and Lord Rayleigh.

University College Bristol, was founded in 1878; Mason College, Birmingham, in 1880; University College, Liverpool, in 1884 and Yorkshire College, Leeds, in 1887.

In 1880 a charter was given to Victoria University (only Owens College at first, later incorporating the Liverpool and Leeds Colleges) and from the very beginning it awarded science degrees. These newer universities were soon playing an important part in the scientific movement of the country.

The University of London admitted women to full degrees in 1879 and the newer provincial institutions such as Victoria University and the Newcastle College of Science made no distinction between men and women in admitting students to their courses. By the end of the century women had access to the majority of institutions offering higher education. (It was only in 1948, however, that Cambridge admitted women to degrees on equal terms with men.)

The Scottish Universities

The high standard of physics teaching in the Scottish Universities during the 18th century was not only maintained but surpassed in the 19th in consequence of a long line of brilliant men who filled the Chairs of Natural Philosophy.

Edinburgh University

In 1819 Sir John Leslie (1766–1832), inventor of his 'Cube' and of the differential air thermometer, succeeded to the Chair of Natural Philosophy. In 1829 the second edition of his *Elements of Natural Philosophy* was published for the use of his students. Throughout the textbook the experimental method is emphasized. Leslie took every opportunity of illustrating what he taught in mechanics, electricity, galvanism, magnetism, optics and astronomy, by means of experiments, using the 'extensive and elegant apparatus' attached to the natural philosophy class. While Leslie was giving the appropriate explanation, an assistant lecturer, trained to the business, would perform the actual experimental demonstration. It is reported that nearly a thousand experiments were demonstrated in a session. The average number attending Leslie's class was between 150 and 160. In 1826 Leslie gave a popular course of physics in the university during the summer session to mixed classes of ladies and gentlemen. But as 'unseemly results' are said to have followed, the innovation was not repeated.

When Leslie died in 1832, he was succeeded in the Chair by a former pupil, James David Forbes. Professor Tait has given us a good idea of Forbes as a lecturer in these recollections:

> His clear, cold, unimpassioned style suited admirably the eternal verity of the laws he enunciated, explained, and illustrated by well-chosen and invariably successful experiments. From the students' point of view he was regarded as too strict a disciplinarian, visiting with what we looked upon as uncalled-for severity very slight infractions of order. This, however, was but the natural outward expression of his own intense earnestness of purpose and sense of duty. All of us who came to know him well found underlying it a grand substratum of geniality and kindly interest.[37]

One of Forbes's students was the young Clerk Maxwell.

When Forbes resigned the Chair of Natural Philosophy in 1860 Clerk Maxwell and P.G. Tait were the candidates. Tait was appointed. In his evidence to the Royal Commission on Scientific Instruction and the Advancement of Science (1874) Professor Tait gives a detailed and illuminating picture of the physics teaching he organized at Edinburgh University, describing at length the topics of physics that he taught, the methods of teaching he adopted, the examinations he set and how he organized the practical work of his students. He sets out clearly the difficulties he encountered from lack of space and inadequacy of funds for apparatus. His Chair was anything but a sinecure; he worked long and exhausting hours, sometimes not even having the help of his 80-year old mechanical assistant. He gave five lectures a week in natural philosophy to the ordinary class and three lectures per week to the advanced class, whilst laboratory work involved 25 hours a week.

> I have two classes to which I lecture. To the first I give over 100 lectures on the whole subject of natural philosophy, carefully explaining and exhibiting the experimental foundations of the subject, and the connexion of the various branches; that is the junior class, and numbers somewhere about 175 students on the average. My second course is to a higher class of students, or students who have attended the junior class for the two or three years preceding. To that class I give three lectures a week; so that in the course of the winter session I give more than 60 lectures to it. Those lectures are upon the applications of the higher mathematics to physical science, treating generally the whole subject of dynamics, the mathematical theory of heat; not only the theory of heat as it was understood by Fourier (the conduction of heat, etc.), but the modern dynamical theory of heat, and also the theory of electricity. In addition to those Sets of lectures which are given by myself, there are tutorial lectures by my class assistant. Those tutorial lectures are given whenever I find a part of a subject which I have not time to develop myself, and when he can be for the purpose spared from his other occupations in the laboratory.
>
> I hold fortnightly examinations of the junior class. Those examinations are introduced as much with a view of ascertaining their proficiency; in fact I find that the use of examinations, not so much for the purpose of giving marks, or anything of that kind, but of finding out the points which the students have not properly understood, is absolutely essential in connexion with teaching; in short I try to find out how far they understand, in order that I may know what I must teach them next, or whether I must go back again.
>
> It is quite impossible to examine so large a class orally without taking a whole day to do it, and therefore what I do is this: I prepare a paper containing about 12 questions entirely upon the subjects which have already been to the best of my ability thoroughly explained in lecture. Those questions are given out orally by me at intervals of five minutes each, and I expect an answer to each in a single short, clear, written sentence. I always endeavour to frame the question so that it admits of an answer in that form, and I refuse altogether to look at or correct, or give any credit for answers of the nature of essays ...
>
> Now I come to another and not less important part of teaching in Edinburgh, and that is the physical laboratory ... It is open at least five days a week, from about 10 to 3 o'clock each day.

> I have made the laboratory open to all comers, limited of course by the number of students which my assistant and I can look after, and which my space can accommodate ... It is impossible for me to accommodate more than 10 or 12 at a time in the space at my disposal ...
>
> I may mention with regard to the teaching in the laboratory, that with the help of my assistant I put each student as he enters the laboratory through an elementary course of the application of the various physical instruments, the primary ones. For instance, I begin by practising them in measuring time, estimating small intervals of time, then measuring very carefully by the micrometer and spherometer, length, angle, curvature; and by other appropriate instruments, temperature, electric current, electric potential and so on ...
>
> Where I find that they have sufficiently mastered those elementary parts of the subject, I allow them to choose the particular branch of natural philosophy to which they wish to devote themselves.[38]

Tait, who was the co-author with Sir William Thomson (Lord Kelvin) of the celebrated Thomson-Tait *Treatise on Natural Philosophy*, was succeeded by Professor Macgregor in 1891. Macgregor was evidently infected by Armstrong's 'heurism'. 'He encouraged his students to become investigators and set them in his laboratory to find things out for themselves ... the only way of getting out of science study the great educational benefit it was capable of affording was to study it by research.'[39]

Glasgow University

The Chair of Natural Philosophy at Glasgow University was held by William Meiklehame from 1803 till his death in 1846. Meiklehame not only gave a lecture every day but also devoted an hour for going over his lecture with his pupils. Three times a week a third hour was devoted to a course of experiments. However, ill-health prevented Meiklehame from teaching much after 1838. Dr John Nichol (1804–59), the Professor of Astronomy, took over Meiklehame's classes in 1840. From 1842 to 1845 Professor David Thomson (1817–80), took over the classes.

In 1846 William Thomson (1824–1907), the future Lord Kelvin, was unanimously elected to the Chair. Thomson had entered Glasgow University in 1834 when just over 10 years old. He was most impressed by Dr John Nichol's course on natural philosophy. After a brilliant career at Cambridge Thomson returned to Glasgow, at the age of 22, to become the most famous professor at Glasgow during the century.

When he arrived he found that the subjects of dynamics and of electricity had been well taught by his predecessors, albeit with apparatus that was quite inadequate. Some of the existing apparatus was almost one hundred years old. There was no provision for the professor to carry out his own research and there certainly never had been any ideas as to providing for laboratory work by the students themselves.

In 1855 Thomson obtained £100 from the University for apparatus and with this and the aid of a band of volunteer helpers from among his students he converted a disused wine cellar in what had been a Professor's house into a laboratory, or experimental room, in which he and his students could work. This was one of the first physics laboratories for students existing in any university – but with a difference. Laboratory work was not compulsory for the students; there was no laboratory course as such; students were free to use the laboratory or not; there were no special hours for attending the laboratory; there was no apparatus specially set aside for student use; there were no demonstrators around to give advice; there were no marks for practical work, and there was no idea of charging a fee. The laboratory was not used for systematic instruction but rather for work by a 'volunteer corps' of students helping Thomson in his researches. This early practical physics teaching was more in the nature of an apprenticeship system – where one or two students learnt their 'trade' at the feet of a master. In course of time, however, the increasing number of students stimulated the teaching of experimental physics in classes rather than to individuals. In 1870 a new laboratory was built.

In his first five years Thomson was allowed £550 for the purchase of new apparatus. This enabled him to replace out-of-date apparatus including apparatus for research in electromagnetism. Many of Thomson's students had the signal privilege of assisting him in his research work. David Murray, one of his students, recalls that Thomson 'had none of the air or manner of a superior, but treated us as if we had been Faradays or Joules'.[40]

In his *Lord Kelvin, as Professor in the Old College of Glasgow* (1924) and in his *Memories of the Old College of Glasgow* (1927) Murray gives us an intimate picture of the great teacher and of how Thomson went about his teaching.

> Lord Kelvin, or as he was in my day, William Thomson, was certainly not a coach, but he was an enthusiastic and inspiring teacher; he aroused and sustained the intelligent interest of his students and gave them a broad and comprehensive view of many branches of physics and awakened in them a love for the subject. No one could listen to him without being imbued with his spirit and being borne along the path he was travelling. Examinations of the present type, except for the degree of M.A. were then unknown. To have converted Professor Thomson into a grinder to cram students for examinations, to coach for honours, to train apprentices in laboratory processes, to set reams of examination papers and read the answers would have blighted his genius and would have ruined him as a scientist ...
>
> He dispensed with the professorial gown after his introductory lectures and stood when lecturing, eager, alert and animated like a runner waiting for the starting signal. He spoke somewhat rapidly, but clearly and distinctly ...
>
> Lord Kelvin possessed the gift of lucid exposition in ordinary language remarkably free from technicalities. ... At the same time he never overlooked elementary principles, and these he repeated and again enunciated, as they emerged in the discussion of almost every topic he touched upon. He dictated a long series of formulae and propositions, expressing the hope that his students would never forget these whatever might be their pursuits in after life. He had no syllabus of lectures and used no notes in lecturing. He had his subject clearly before him and dealt with it in logical order ...
>
> In this way the Professor found time to deal very thoroughly with general physics and with special branches such as heat and electricity. It was, however, rather hard upon the students, they had to work privately at Dynamics and at the same time keep pace with the Professor's lectures, to say nothing of their reading on Astronomy. Fortunately for us his rate of speed was occasionally interrupted. A thunderstorm, for instance, was a godsend. The electrometers were all at work,

observations taken, explanations given and a running commentary made; conjectures thrown out and the whole treated as an experiment specially designed to illustrate some point which had been touched upon in recent lectures. It was not an observation of ascertained phenomena but an opportunity for collecting fresh data for further investigation. Everything which was observed was explained, discussed and commented on, and this commentary might be continued on the next day or even longer as fresh suggestions occurred.

When the professor got beyond his class, he recognised this and would pause and say that they must assume certain propositions or allow him to work out certain calculations ... An important element in his method of teaching was that it was always up to date. Each topic was presented in its latest phase, and in so presenting it he often carried it further than had already been done and made suggestions beyond those that had been published.[41]

This last sentence is confirmed by one of the questions set in the degree examination in April 1848. 'State briefly Carnot's principle according to which the motive power of heat must be estimated. Explain a principle according to which from this theory an absolute thermometer scale may be defined.' It was only two weeks earlier that Thomson had read a paper to the Philosophical Society of Glasgow 'On an Absolute Thermometer Scale Founded on Carnot's Theory'.

Two foreign observers – Demogest and Montucci – have recorded a description of a visit they made in 1869 to one of Sir William Thomson's natural philosophy classes.

Surrounded by 60 students he explained to them, by the aid of experiments, the theory of capillary attraction. As he proceeded he questioned some students in the most affable manner so that there sometimes resulted a short scientific conversation. An adjoining room was reserved for students who wished to practise handling the apparatus and to repeat experiments already seen.[42]

James Coutts also gives us an intimate account of Thomson's teaching in his *History of the University of Glasgow* (1909).

Some other experiments were great annual events in the class, such as the firing of the shot. A log of wood was slung up over the platform in such a way as to swing backward and forward with the impulse imparted to it by a bullet, and there was a contrivance to fix the log at the limit of its swing, so as to show the distance it had moved, the movement being a rough equivalent to the explosive force of the gunpowder ... It was a sight to be remembered to see the amused curiosity with which the students watched him seizing his muzzle-loading rifle, cocking the doghead, seating himself on a front bench, placing the butt-end of the weapon to his shoulders, taking cool and resolute aim through his spectacles, and lastly firing the shot. This demonstration was wisely reserved for the end of the hour, for it would have required a good while to restore the students to gravity and order.[43]

Thomson was knighted in 1866 for his share in the successful laying of the Atlantic Cable. In 1892 he was created a peer – Lord Kelvin. In 1899 he retired from the Chair of Natural Philosophy and was succeeded by Professor Andrew Gray.

Aberdeen University

Aberdeen University comprised King's College, Aberdeen, and Marischal College. During the first half of the 19th century natural philosophy formed part of the third-year course in both colleges.

At King's College the natural philosophy course in 1831 covered general physics, dynamics, mechanics, hydrostatics, pneumatics, acoustics, optics, electricity and magnetism. At Marischal College natural philosophy was made compulsory for the Arts degree in 1826. The course embraced dynamics, the theory and practice of machinery, electricity, magnetism, optics and astronomy. Two rooms were set apart for the natural philosophy apparatus, which is said to have been extensive.

James Clerk Maxwell was appointed Professor of Natural Philosophy at Marischal College in 1856. Maxwell spent four years at Marischal College. One of his students there, David Gill, has left this charming picture of the man whose teaching 'influenced the whole of my future life'.

> In those days there was no working physical laboratory at Aberdeen accessible to students, but simply an apparatus room containing the old-fashioned lecture models of levers, pulleys, pumps, windmills, steam engines, etc., with some balances, air pumps, tuning forks, and Atwood's machine, electric machines, a few galvanic batteries, etc., in glass cases; but access to this room was forbidden to students. After the lectures, however, Clerk Maxwell used to remain in the lecture-room for hours, with some three or four of us who desired to ask questions or discuss any points suggested by himself or by ourselves, and would show us models of apparatus he had contrived and was experimenting with at the time, such as his precessional top, colour box, etc. These were hours of purest delight to me. Maxwell's lectures were, as a rule, most carefully arranged and written out – practically in a form fit for printing – and we were allowed to copy them. In lecturing he would begin reading his manuscript, but at the end of five minutes or so he would stop, remarking, 'Perhaps I might explain this,' and then he would run off after some idea which had just flashed upon his mind, thinking aloud as he covered the blackboard with figures and symbols, and generally outrunning the comprehension of the best of us. Then he would return to his manuscript, but by this time the lecture hour was nearly over and the remainder of the subject was dropped or carried over to another day. Perhaps there were a few experimental illustrations – and they very often failed – and to many it seemed that Clerk Maxwell was not a very good professor. But to those who could catch a few of the sparks that flashed as he thought aloud at the blackboard in lecture, or when he twinkled with wit and suggestion in after lecture conversation, Maxwell was supreme as an inspiration.[44]

Chapter Fifteen

The Cavendish Laboratory

James Clerk Maxwell (1831–79)

In 1852 the Commissioners looking into the state of the University of Cambridge expressed the view that experimental science might help to solve the student problem of the time. The university was threatened by a dangerous class of undergraduates who were exposed to the temptation of finding 'mean and frivolous ways in which wealth may be squandered and leisure abused'. The Commission recommended that the university should 'teach the handling of meteorological and magnetic instruments, and cameras'. If this were done some of them might 'become Humboldts'.

In 1868 the scheme of examination for honours in the mathematical tripos was extended to include the subjects of heat, electricity and magnetism. A syndicate appointed in November of that year to consider how these subjects might best be taught recommended the founding of a special Professorship of Experimental Physics and of supplying the Professor with the means of making his teaching practical. This implied the further provision of a demonstrator, of a lecture room, a laboratory and several classrooms, with a sufficient stock of apparatus. The cost of the buildings and apparatus was estimated at £6300. But where were the funds to come from? In October 1870 the Duke of Devonshire, who was Chancellor of the University, offered to provide the funds. As a member of the Royal Commission on Scientific Education he had perceived how useful such an institution might become.

Who was to be the first holder of this Chair of Experimental Physics? It was understood that Sir William Thomson had declined to stand. Attention turned to James Clerk Maxwell. His researches had placed Maxwell in the forefront of physical experimenters.

In February 1871 the Hon. J.W. Strutt, the son of Lord Rayleigh, wrote to Maxwell from Cambridge:

> When I came here last Friday I found everyone talking about the new professorship, and hoping that you would come. Thomson, it seems, has definitely declined. There is no one here in the least fit for the post. What is wanted by most who know anything about it is not so much a lecturer as a mathematician who has actual experience in experimenting, and who might direct the energies of the younger Fellows and bachelors into a proper channel. There must be many who would be willing to work under a competent man, and who, while learning themselves, would materially assist him ... I hope you may be induced to come.

Another to press Maxwell to stand was Professor G.G. Stokes.

Maxwell was elected, unopposed, to the Chair on 8 March 1871 'to teach and illustrate the laws of Heat, Electricity and Magnetism'. In October of that year Maxwell began his course of lectures on the subjects prescribed in his commission. At the beginning, the lectures were delivered to very small audiences, which was just as well, since there was nowhere to accommodate more. In October 1872 Maxwell wrote: 'The Laboratory is rising, I hear, but I have no place to erect my chair, but move about like the cuckoo, depositing my notions in the Chemical Lecture Room in the first term, in the Botanical in Lent and in Comparative Anatomy in Easter.'[1]

For some time after his appointment Maxwell's main work was that of designing and superintending the erection of the Cavendish Laboratory. Maxwell, accompanied by Mr Trotter, visited various laboratories at home and abroad to gain ideas. The selection and design of apparatus and equipment for the Laboratory also occupied a good deal of Maxwell's time. The building of the laboratory cost far more than the original estimate, but the Chancellor's generosity was not limited and he wrote to the Vice-Chancellor: 'It is my wish to provide all the instruments for the Cavendish Laboratory which Professor Maxwell may consider to be immediately required, either in his lectures or otherwise.'

The laboratory was completed in 1874. It was named after Henry Cavendish (1731–1810), who belonged to one of the oldest and richest families in England. He lived in London, in solitude and great retirement, devoted entirely to science. He did original and important work on the combination of gases, such as hydrogen and oxygen, and on electrical measurements; he measured the density of the earth to a value not very different from today's measurement.

In his fourth report to the syndicate Maxwell declared that the laboratory's apparatus was complete, comprising 'the whole of the instruments required by the prominent state of science' – a happy occasion that has never been attained again! Among this apparatus were a tangent galvanometer, a discharging key, apparatus for measuring the viscosity of gases, for colour mixture, a dynamical top, a model of Saturn's rings and a collection of apparatus which had belonged to Professor Wollaston.

During Maxwell's professorship the number of students studying experimental physics was small. The mathematical tripos tradition was still very strong, and even though the physicists had succeeded in getting questions on heat, electricity and magnetism introduced into the tripos examinations in the new regulations for the examinations in 1873, few students felt it worth while spending time on experimental physics. Experimental work was thought hardly fitting for wranglers. Maxwell was brought in as an additional examiner, especially for the physical subjects.

A laboratory examination in practical physics was introduced in a tentative way into the natural science tripos in 1874. Candidates were presented with questions on chemistry, mineralogy and physics. Maxwell was the examiner for physics. The physics involved such questions as determining the focal length of a lens, the time of vibration of a magnet, or the electrical resistance of a wire. In the following year the natural science tripos was divided into two parts, taken in June and December respectively. In the second part a separate paper was set, containing eight laboratory questions, of which not more than four were to be attempted.

The demonstrator William Garnett, a former student of Maxwell, superintended the ordinary laboratory practice, though Maxwell himself frequently instructed students, especially those who were intending to undertake research work after having gained preliminary experience in physical measurements. On the question of Laboratory teaching Garnett writes:

> Maxwell's idea at first was to attach to the Laboratory a small band of graduates, each of whom would undertake a definite piece of work after going through a short course of training in measurement. For its educational value he held in high repute the Kew Magnetometer, as it afforded practice not only in reading scales and making adjustments, but also in time observations, counting the beats of a watch, while observing the vibrating magnet. At first all students coming to the Laboratory were invited to measure the horizontal component of the earth's horizontal magnetic force, making their own calculations before proceeding to any other piece of work. It was not then contemplated that the Laboratory would be much used by undergraduates [2]

The distinction of being the first student to work in the Cavendish Laboratory belongs to W.M. Hicks, of St John's College, who later became Professor of Physics at the University of Sheffield. Hicks gives this account of his work at the laboratory.

> Beyond resistance measurements and using the Kew Magnetometer to get H with all corrections and under the best conditions, I do not remember doing any of the 'regular' things. When Lippmann published the principle of his electrometer, I designed one on the basis, and made one, and the getting that to act properly was worth any amount of routine measurements. As illustrating the effect of Maxwell's teaching, I was fired with the desire of measuring experimentally the velocity of propagation of electromagnetic waves (1874), and designed a piece of apparatus to prove its existence, depending on the action of two coils, a large one and a very small one, on the same circuit, acting on a light magnetic needle at a point where their combined force was zero. As the large coil was much further away, I supposed that when the current was suddenly started – or suddenly stopped – the effect of one at the needle would disappear before that of the other, and the needle would give a kick. Of course nothing came of it, but the practice was worth a great deal to me, and it is interesting as showing the kind of atmosphere in which one was working round Maxwell.[3]

J.E.H. Gordon of Caius College was the next to join the Laboratory. Under Maxwell's guidance he undertook research to determine accurately some electrical constants.

Another early worker at the Cavendish was George Chrystal (Second Wrangler in 1875). Following doubts expressed about the validity of Ohm's Law at a British Association meeting in 1874 Chrystal showed from accurate measurements that the Law was more than a rough approximation. Commenting on Chrystal's confirmation of the accuracy of Ohm's Law, Maxwell stated: 'the simplicity of an experimental law may be an argument for its exactness.' J.A. Fleming, later to be Pender Professor of Electrical Engineering, gave up his science mastership at Cheltenham College to work under Maxwell. He continued Chrystal's experiments of comparing the BA units of resistance and determined the most probable value of the ohm.

In a letter to his friend Lewis Campbell in Christmas 1876 Maxwell wrote:

> As we get richer in apparatus, mathematical lectures give way to experimental, and the blackboard to the lamp and the scale. I have had a pupil quite innocent of mathematics who has learned to measure the focal length of lenses, and has found the e.m.f. from the water-pipes to the gas-pipes, and from either set of pipes to the lightning conductor ....
>
> I have been making a mechanical model of an induction coil, in which the primary and secondary currents are represented by the motion of wheels, and in which I can symbolise all the effects of putting in more or less of the iron core, or more or less resistance and Leyden jars in either circuit.
>
> I have also been making a clay model of Professor W. Gibbs' thermodynamic surface, representing the relations of the solid, liquid and gaseous states, and the different paths by which a body may get from the one to the other.[4]

A.P. Trotter, in a letter to Schuster, provides some idea of the sort of instruments that were available to students in the laboratory during their course of instruction at that time.

> Either I or someone else made a light-hearted attempt at measuring the current of the Holtz machine by means of a high-resistance Elliott galvanometer. Somebody scolded the culprit, but I could not see the enormity of the offence. Among the apparatus that I can remember using was a small spectroscope with which I measured the angles and refractive index of a prism. I had a book of trigonometrical tables in my hands for the first time. I knew all that the extras of the Little-go required as to trigonometry, but I could not understand why the sines and cosines stopped at 45°. I wanted an angle of about 60° and thought it must be in a second volume! I remember your allowing me to see the double sodium line in your big spectroscope, as a treat. I spent several days with the Kew magnetometer, and Garnett explained how to calculate the moment of inertia of the magnet – a quantity of which I had never heard.[5]

In 1876 a course of lectures in heat and light for Part I of the natural sciences tripos was announced with the following disarming notice: 'In these lectures no previous mathematical knowledge beyond Arithmetic will be assumed.' As the syllabus included diffraction fringes, gratings, circular and elliptical polarization and transmission of light through crystals, it would be interesting to know how far this limitation was in point of fact achieved. In 1877 this lecture course was thrown open to medical students and other undergraduates. These lectures soon became popular and found some of the largest science classes in the university. So large, in fact, that Maxwell's successor – Lord Rayleigh – abolished practical physics lectures to the medical students as taking up too much of the time and resources of the Cavendish.

In 1877–8 the regulations of the mathematical tripos were altered in a way which tended to discourage more students from doing physics. Physics was removed from the first part of the tripos and combined with the more advanced parts of astronomy and pure mathematics. Under this arrangement very few students chose to read physics. However, in 1879 a new course was introduced in the long vacation for more systematic instruction in practical work.

For several years Maxwell refused to open his laboratory to women. When at last he consented women were admitted while he was away on long vacation in Scotland. Garnett taught a class of ladies who were determined to go through a complete course in electrical measurement.

Arthur Schuster, later Professor of Physics at the University of Manchester, began his spectroscopic researches at the Cavendish in 1876. He records that students were left largely to get on with things themselves. He contrasted this with the 'over-instruction' which was to become common in many laboratories later. Schuster describes how Maxwell's personality dominated the Cavendish and united those who worked in it.

> Maxwell took an active part in every research that was carried out in the laboratory. He used to come round daily, enter the various rooms, and say a few words, discussing results or making suggestions ... He was always encouraging, even when he thought a student was on the wrong track. 'I never try to dissuade a man from trying an experiment,' he once told me. 'If he does not find what he wants, he may find out something else' ....
>
> He often showed a certain absent-mindedness: a question put to him might remain unnoticed, or be answered by a remark which had no obvious connection with it. But it happened more than once that on the following day he would at once refer to the question in a manner which showed that he had spent some time and thought on it .... It was quite usual for him to begin a conversation with the remark: 'You asked me a question the other day, and I have been thinking about it'.[6]

In 1878 James Ambrose Fleming attended Maxwell's lectures on electricity and magnetism which he described as 'splendid'. Fleming confirms how good Maxwell's lectures were and how little they were appreciated.

> Maxwell gave a couple of courses of University lectures per year, which, however, were shockingly neglected, and I remember my surprise at finding a teacher whom I regarded as in the very forefront of knowledge lecturing to three or four students as his only audience. In fact during one term Professor Maxwell gave a course of splendid lectures on Electrodynamics, the only audiences being myself and another gentleman whose name I think was Middleton, at that time a fellow commoner of St John's.[7]

Another student has recorded that 'Maxwell's lectures were the delight of those who could follow them in his brilliant expositions and rapid changes of thought.'

Maxwell's dog, Toby, accompanied him everywhere, even into the laboratory. Everitt writes that Toby was thoroughly conversant with the details of the laboratory and some of its apparatus. He always betrayed signs of uneasiness when he heard electrical sparks, but when summoned to his post he would sit down between his master's feet and allow the electrophorus to be excited upon his back, growling all the time in a peculiar manner, as though to relieve his mind, but not evidencing any signs of real discomfort.[8]

In 1878 J.H. Poynting resigned his post as demonstrator to Professor Balfour Stewart at Owens College, Manchester, to continue his measurements to determine the gravitational constant by weighing at the Cavendish. In a letter to Joule Maxwell wrote, concerning Poynting's work, 'You see that the age of heroic experiments is not yet past.'

Maxwell himself did very little experimental work on his own in the Cavendish. This was partly due to the claims of his teaching work, partly to the personal interest he took in the experimental work of others, and partly also to the fact that his most active interests at the time were theoretical; but it was largely due to the task which he assumed of editing the unpublished electrical writings of Henry Cavendish (1731–1810). (In Cavendish's day there was no means of measuring the strength of an electrical current. Cavendish used to pass an electric current through his body and estimate its magnitude by the intensity of the resulting shock!)

A tribute to Maxwell's stewardship of the Cavendish Laboratory was paid by Sir Arthur Shipley in his reminiscences: 'In the following year [i.e. 1880] when I came up to Cambridge ... the Cavendish Laboratory was then, as now, doing work which has made it the Mecca of all physicists throughout the world.'[9] Writing from the University of Glasgow on 21 January 1882, Sir William Thomson wrote: 'There is, indeed, nothing short of a revival of Physical Science at Cambridge within the last 15 years, and this is largely due to Maxwell's influence.'[10]

Lord Rayleigh

Maxwell died in November 1879. Lord Rayleigh was invited to apply for the Chair and although he valued his freedom and independence as a landowner, financial difficulties at the time inclined him to accept. On 12 December 1879 he was duly elected.

Lord Rayleigh writes that before coming to Cambridge to take up the Chair he had not given any particular attention to the subject of electricity. But he found Cambridge to be so saturated with the subject that he quickly came to the conclusion that it would be best to make it his particular study. 'All this was owing to the influence of Maxwell.'

Rayleigh commenced work in the Lent Term of 1880, lecturing on the use of physical apparatus. He was required to reside in Cambridge for 18 weeks during the academic year and to give one course of lectures in each of two terms, 40 lectures at least, during the year. During the year R.T. Glazebrook and W.N. Shaw were appointed demonstrators of physics and they began regular courses of instruction in practical physics. The notice for Michaelmas Term 1880 announces that:

> The Professor of Experimental Physics will lecture on Galvanic Electricity and Electromagnetism. Dr Schuster will give a weekly course of demonstrations on Electricity and Magnetism, Mr Shaw will give a course of demonstrations on the Principles of Measurement and the Physical Properties of Bodies.
>
> The Laboratory will be open daily from 10 to 4 for more advanced practical work under the supervision of the Professor and Demonstrators for those who have had the necessary training. Courses of demonstrations will be given during the Lent Term on Heat and Advanced Electricity and Magnetism, during the Easter Term on Light, Elasticity, and Sound.

These various courses of demonstrations were each for two hours, three days a week. The number of students attending the lectures and demonstrations at first was small. In the Lent Term, 1880, 16 attended Lord Rayleigh's lectures, 18 the elementary demonstrations on heat, and 14 the advanced demonstrations on electricity and magnetism.

In 1881 Rayleigh secured the services of an efficient mechanic, George Gordon, and so began the workshop, which in later years became so important a part of the Cavendish, for the construction and repair of instruments for lecture demonstrations, for the practical classes and for research.

Practical experimental work was not made a compulsory part of the education of physics undergraduates at the Cavendish Laboratory until Rayleigh arrived. Lord Rayleigh aimed at developing the systematic teaching of elementary practical physics to a large number of students. Such practical work demands large numbers of apparatus, such as galvanometers, etc. Inexpensive, simple, standardized instruments, manufactured in quantity, were not then available for the simple reason there was no demand for them. In fact, on taking charge of the Cavendish Laboratory, Rayleigh was struck with the inadequacy of the available apparatus. In his first annual report he complained, in particular, that 'there is no steam engine or other prime mover; nor among the acoustical apparatus is there any musical instrument'. To remedy this situation he inaugurated an 'Apparatus Fund' to which he himself contributed £500.

Glazebrook points out how the teaching of practical physics in those days was an undeveloped art. The time given to practical work was not long and there was comparatively little systematic endeavour to teach the fundamental physical principles to the ordinary student by means of experiments which he performed himself. Nor were there textbooks on practical physics to help the student.

Professor Pickering of the Massachusetts Institute of Technology had recognized the demand for a special textbook of practical physics for use by students and accordingly wrote one – his *Elements of Physical Manipulation* – in 1873. In the book he outlined how he taught classes of practical physics.

> The method of conducting a Physical Laboratory for which this book is especially designed and which has been in daily use with entire success at the Institute is as follows. Each experiment is assigned to a table on which the necessary apparatus is kept and where it is always used. A board called an indicator is hung on the wall of the room, and carries two sets of cards opposite to each other, the one bearing the names of the experiments, the other those of the students. When the class enters the laboratory each member goes to the indicator, sees what experiment is assigned to him, then to the proper table where he finds the instruments required, and with the aid of the book performs the experiment.
>
> By following this plan an instructor can readily superintend classes of about 20 at a time and is free to pass from one to another answering questions and seeing that no mistakes are made.[11]

Most physicists will be familiar with this plan of practical physics teaching. Much of the work of Lord Rayleigh and his staff was devoted to the introduction of this system of practical teaching at the Cavendish Laboratory.

Since the textbook by Pickering was not entirely suitable for the Cavendish, Glazebrook and Shaw prepared a series of MS notebooks, each dealing with one experiment, describing the apparatus actually in use in the laboratory, the method of marking the experiments and in particular of entering up the results. The note books were eventually published as *Practical Physics.* This covered the ground of the elementary portion of the work required for the first part of the natural science tripos.

The advanced classes were worked in a somewhat similar way, but from the beginning an attempt was made to introduce students to original work and important researches by endeavouring to repeat such of these as were possible with the means at their disposal. Students were expected to read the memoirs referred to and in their work to reach a high order of accuracy.

Glazebrook provides this personal record of teaching at the Cavendish under Lord Rayleigh.

> The course was planned out and the syllabus was arranged with him after most careful consideration and discussion; otherwise we were free to teach as seemed best and to develop the work in those directions which appeared to us to be most necessary.
>
> Each of the two Demonstrators was on duty three days a week, and during those three days there was not much time for anything besides the work of the class. On the other three days we were free for research work, College duties, or private teaching, as the case might be. The conditions of work were delightful. The pleasure of developing a comparatively new line of teaching, the sympathy and support of the Professor, the interest in seeing the work grow and the realisation of the importance of the fruits that were to come, all combined to make those years 1880–1884 a period of real happiness to the staff of the Laboratory.[12]

The most marked result of the Rayleigh period at the Cavendish were his own splendid researches. Rayleigh's method of work was to choose a subject of research which could be planned on a large scale so that several workers could share in the work. He chose the accurate re-determination of electrical units which at that time were uncertain by as much as three or four per cent. Rayleigh's team made it possible to measure to a high degree of accuracy an electric current in amperes, an electromotive force (e.m.f.) in volts, and a resistance in ohms. In 1894 the Board of Trade adopted these measurements as the legal standards of electrical units. Rayleigh's pursuit of Maxwell's principle of continually increasing the accuracy of experiments as a means to revealing new and unexpected facts led him to his discovery of argon and the rare gases.

Among other researches that came from the Cavendish Laboratory during Rayleigh's tenure were those of Glazebrook on optics, J.H. Poynting on the saccharimeter, J.J. Thomson on the ratio of electrical units, Horace and George Darwin on gravitation and W.N. Shaw on meteorology. When Rayleigh left the Cavendish in 1884 on his appointment as Professor at the Royal Institution it was firmly established as a leading centre of learning and research.

J.J. Thomson

When Lord Rayleigh resigned the Chair of Experimental Physics in 1884 several candidates applied for the post. Among these were the two able demonstrators Glazebrook and Shaw, Schuster, and Osborne Reynolds, who was Professor of Engineering at Manchester. Sir William Thomson was approached but he declined. On 22 December 1884 it was announced that Joseph John Thomson had been elected. The news came as a great surprise. Thomson was only 28 and his preferment over able and experienced physicists, many of whom had been his teachers, was completely unexpected – most so by J.J. himself.

> To my great surprise, and, I think, to that of the rest of the University, I was chosen. I remember I was told at the time that one well-known college tutor had expressed the opinion that things had come to a pretty pass in the University when mere boys were made professors.'[13]

He described himself as feeling like a fisherman with a light tackle who had casually thrown in his line at an unlikely spot and found he had hooked a very large fish too heavy to land.

The strain on Thomson at the beginning of his professorship was great. Besides having to adjust to the psychological problems of having to direct students who had been his seniors and teachers and having to cope with the problems of administration and management of which he had no experience, Thomson was profoundly conscious of the intellectual responsibility of succeeding men like Maxwell and Rayleigh. When he gave his first course of lectures as Cavendish Professor in 1885 he nearly fainted and had to dismiss the class.

Thomson followed the practice of Rayleigh giving two courses in a year, one on properties of matter and the other on electricity and magnetism. They were not designed as preparatory for the tripos examinations but to be generally stimulating; and they drew large audiences.

By 1885 the number of students attending demonstrations had increased to 90 and about 10 students were engaged in original research. In due course Thomson's lectures came to be very much appreciated by his students, whether as guidance in reading for the tripos or for suggestions of research in postgraduate study. H.F. Newall, one of Thomson's early assistants, recalls Thomson's

> power of suddenly introducing an old friend in a new light, his patience in remembering that many of his learners had not had a sympathetic training in mathematics and would always be the better for being led from first principles to the point of view he wished to emphasize, these were qualities that made his lectures supremely valuable to his students both in mathematical and experimental physics ...
>
> J.J. was very awkward with his fingers, and I found it necessary not to encourage him to handle the instruments ... On the other hand, his real interest was experimental ... His intuitive ability to comprehend the inner working of

intricate apparatus without the trouble of handling it, appeared to me then, and still appears to me now, as something verging on the miraculous, the hallmark of a great genius.

Thomson's research lay in the conduction of electricity through gases. In 1887 Hertz detected electromagnetic waves and so provided experimental proof of Maxwell's Electromagnetic Theory. Thomson repeated these experiments in his classes in 1888. In the meanwhile a new group of classes in practical physics began to develop in connection with the Elementary Lectures. The number of medical students attending had increased steadily each year and now formed a large majority of the class. In the examination for the medical bachelorship candidates were set written questions only. In 1887 a simple oral examination on the use of practical apparatus was introduced – 'with spectacular results' as Crowther remarks. 'The memories of the first examiners almost collapsed under the strain of storing for future recital the immense collection of "howlers" perpetrated by their victims.'[14]

In 1887 T.C. Fitzpatrick was appointed as assistant demonstrator to organize a class in elementary practical physics for medical students in connection with Glazebrook's lectures. The classes grew rapidly and further assistant demonstrators had to be appointed. As the laboratory space by now was overcrowded a corrugated iron shed which had been used as a dissecting room by the Department of Anatomy and had become vacant was used for the medical students' practical physics in 1890.

In 1892 the regulations of the mathematical tripos were changed. The subjects of heat, elementary hydrodynamics and sound, along with the simpler parts of electricity and magnetism were included in Part I of the examination. The more advanced parts of mathematics and physics were included in Part II. Whilst in 1874 only 17 men took the natural sciences tripos and few of these did physics, 20 years later the numbers had much increased. In 1894 103 candidates took Part I of this tripos and 16 Part II.

Also in 1894, provision was made whereby students who had graduated at other universities could come to Cambridge as 'research students' and take a degree after two years' advanced study and research. The first student to take advantage of this scheme was the New Zealander Ernest Rutherford, who was later to succeed J.J. Thomson as Cavendish Professor. These students were the precursors of a very numerous and distinguished band of students who came to the Cavendish Laboratory from every important university in Europe, America, Asia and Africa.

In 1896 Wilberforce and others who were demonstrators produced a *Laboratory Notebook of Elementary Practical Physics* for the use of the many students in the Cavendish Laboratory. There was a rapid increase in the number of candidates for Part I of the natural sciences tripos and the proportion of students taking physics was well maintained. The elementary classes were very full. A new practical class, under Skinner, was started in connection with lectures of Shaw, who would arrive at the lecture room in morning coat and silk hat. At the same time G.F.C. Searle was organizing elementary classes along the lines originally laid down by Glazebrook and Shaw. In this system it was only possible to have one example of each apparatus and so the order in which the experiments were done did not synchronize with the lectures but depended on what apparatus was not in use. In Skinner's class there were fewer experiments but for each of these several sets of apparatus were needed and so the experiments done synchronized with the lectures.

In addition to his university classes, Searle also organized a class in the long vacation for science masters who wished to keep abreast with developments in the teaching of practical physics, and also for mathematics masters who wished to illustrate their teaching of mechanics by experimental illustrations. The many research students coming to study under J.J. aggravated the shortage of accommodation and of apparatus. It was said that more physics was done in the Cavendish per square centimetre than in any other laboratory in the world. As for apparatus it was often a case of appropriating what you needed when the users were not looking.

In 1904 Lord Rayleigh was awarded the Nobel Prize for physics. He donated £500 of the prize to the Cavendish Laboratory, which went part of the way to build a new Rayleigh Wing of the Laboratory. The Rayleigh Wing was opened in June 1908 and for the first time the whole of the physics lectures in the university were given in the Cavendish Laboratory. The chief function of the college tutors was now to supplement the Cavendish courses and help students who needed individual instruction.

In 1906 Professor Thomson was awarded the Nobel Prize for physics. In 1909 he was knighted. In 1908 it was the turn of Ernest Rutherford to be awarded the Nobel Prize – for chemistry.

Monumental pioneering research work was being done at the Cavendish. C.T.R. Wilson had been working on his cloud chamber from 1897, the perfections of which, in 1911, at a total expenditure of about £5, was to help so much of the fundamental work being done around him. W.L. Bragg, who was to win the Nobel Prize in 1915, entered Cambridge as a student in 1909. He reminisces on Wilson as a teacher there.

> C.T.R. Wilson lectured on optics. The lectures were the best and the delivery was the worst, of any lectures to which I have ever been. He mumbled facing the board, he was very hesitant and jerky in his delivery, and yet the way he presented the subject was quite brilliant. I think his lectures on optics set the standard for similar lectures all over the country when later his pupils got Chairs.[15]

E.V. Appleton, who entered Cambridge as a student in 1911, was fascinated by Wilson's lectures.

> His manner of lecturing was his own. He lectured with both hands, writing on the blackboard with one hand, and wiping off what he had written a few seconds later, with the other. His learners had to be pretty intent to follow what was going on, which they were most anxious to do, knowing that what was being said was of unique quality ... . He didn't want another of the great men of the Cavendish, G.F.C. Searle, to acquire his material for one of his textbooks![16]

Both G.P. Thomson (son of J.J.) and W.L. Bragg gained crucial technical suggestions from Wilson's Cavendish course on optics, the former for his discovery of electron diffraction and the latter for X-ray crystal analysis.

W.L. Bragg, who was himself to become the Cavendish Professor in 1938, returned to Cambridge as a lecturer in 1914. The next year the Nobel Prize was awarded to him and his father (W.H. Bragg) jointly for their work on X-ray crystallography. Not surprisingly X-ray analysis figured large in the research at the Cavendish.

Besides Wilson, others engaged in research at the Cavendish were W. Wetham working on electrolysis, Rutherford on uranium radiation, O.W. Richardson on thermionics., H.A. Wilson on the conductivity of flames. The outbreak of war in 1914 brought this period of research to an end. It marked the end of an era when pioneering work could be done with very simple and inexpensive apparatus – 'with sealing wax and string'. It also marked the end of Thomson's tenure of the Cavendish Chair.

Ernest Rutherford

Thomson resigned as Cavendish Professor of Experimental Physics in 1918. Rutherford stood for election at the last possible moment, actually applying by telegram on the day of the election. Perhaps when one has a Nobel Prize, has fathomed the nature of radioactivity, has established the nuclear theory of the atom, has succeeded in artificially disintegrating the atom (at Manchester in 1917) and has founded nuclear physics, one can take such chances. Rutherford was then Professor of Physics at Manchester, and 47 years old. On 2 April 1918, he was appointed Cavendish Professor in succession to Thomson.

Rutherford, a New Zealander, had gone to the local village school at the age of five. At ten he had acquired a primer of science written by Balfour Stewart (who, as Professor at Manchester, had been J.J.'s teacher). In due course Rutherford entered the Cavendish Laboratory as the first postgraduate research student in 1894.

In his recollections of his Cambridge days the Australian physicist Oliphant (born 1901) gives us some illuminating accounts of his studies under Rutherford.

> After getting entangled in calculations about the radius of the electron, he turned to the class, and told them they could complete it for themselves; in any case it had 'very little meaning'. On another occasion, getting into a contradiction about the apparent increase of mass of an electron with increase in its kinetic energy, he suddenly said to the class: 'You sit there like a lot of numskulls, and not one of you can tell me where I've gone wrong.' Technically, he was not a good lecturer, except when speaking on a recent discovery of his own.[17]

In course of time, because of his outside engagements, Rutherford had to depute his lectures on general atomic physics and Oliphant sometimes did these for him. When this happened Rutherford called him to his room, and went through the lecture he was to give. He had an extraordinary collection of notes, written on scraps of paper, all pinned together in the correct order. He had used these notes for years, altering and adding to them as knowledge and experience advanced. The emendations were written in pencil, often almost illegible. They went through these together for an hour or more, making more alterations, and muddling their order; Oliphant found it necessary to write out a set of notes of his own, and as he received only a day's notice, this was quite a heavy task.

At this time a whole coterie of eminent students were working at the Cavendish, among them G.P. Thomson (J.J.'s son), P.M.S. Blackett, E.V. Appleton (famous for his determination of the height of the ionosphere), and J.D. Cockcroft. A number of major discoveries were made, including the new isotope of hydrogen, mass number 3, which was named tritium and is a constituent of the H-bomb.

Cavendish Professors since 1938

Rutherford died in 1937. **W.L. Bragg** had succeeded Rutherford as Professor at Manchester University in 1919. He now succeeded Rutherford as Cavendish Professor in 1938.

During World War II teaching at the Cavendish still went on. Before the war the annual number of honours graduates in physics was about 160; in 1948 the figure was 500. Compared to about 40 researchers in the Laboratory before the War, there were 160 after the War, divided into six fields – nuclear, low-temperature, radio, crystallography, metal and mathematical physics. The teaching of experimental physics was done in five practical laboratories, each with its own staff. The total staff consisted of three professors, four readers, eight lecturers in experimental and two in theoretical physics. In his last report for 1952–3 Bragg mentioned that the course of arts lectures that had been introduced to widen the outlook of physics students continued to be successful with an average audience of 200. Bragg also favoured a scheme whereby honours physics students were recommended to work for six months in industry before coming to the university.

Bragg was particularly helpful to Ryle who was then trying to fathom some of the mysteries of radio-astronomy. Bragg perceived that the problem of deducing the size and position of radio objects in the sky by interferometric methods was the exact analogue of the determination of the arrangement of atoms in a biological molecule that was engaging the attention of the X-ray analysts Perutz and Kendrew. It was almost incredible that the structure of the cosmic universe and the structure of molecules at the basis of life were being investigated at the same time in the same laboratory by the application of the same optical analytical technique.

Bragg himself was a great teacher and lecturer. He taught teachers how to teach and set up new illustrative experiments. In 1953 Bragg left for the Royal Institution where he inaugurated a system of lectures for schools in which he introduced tens of thousands of schoolchildren to science with unsurpassed skill.

N.F. Mott, who had created a very reputable school of physics at Bristol University, succeeded Bragg as Cavendish Professor in 1954. In 1966 Mott reported an establishment of six professors, 10 readers, 15 university lecturers, seven assistant directors of research, four senior research assistants, nine senior technical officers, six assistant technical officers, and one computing officer, making a total of 74 staff.

In 1971 **A.B. Pippard** succeeded Mott as Cavendish Professor, giving his inaugural lecture almost 100 years to the day of Clerk Maxwell's introductory lecture in October 1871. By 1980–1 the number of professors on the staff of the Cavendish Laboratory had increased to 10.

Chapter Sixteen

Physics in 20th Century Schooling

Elementary schools

In the early years of the century the bulk of elementary school pupils left school at the age of 14 or earlier. The only scientific knowledge they acquired was through a series of object and observation lessons designed to link up facts of everyday life with simple scientific truths. The quality of the work varied widely. In the best elementary schools the object lessons were very good, but in many the lessons were inadequate and often seriously so. Many schools had no laboratories and had little or no science equipment either. Armstrong complained that the conventional object lesson in 1902 provided a model of 'what should be avoided'; and the British Association for the Advancement of Science remarked that by 1908 the object lesson had often deteriorated to become little more than 'a laborious elucidation of the obvious'.

From about 1887 onwards, as a result of the campaign by Armstrong that pupils carry out their own experiments, the theory had developed that no science teaching was worth the name unless it involved pupil experimental activity; demonstrations were regarded as positively dangerous. Experimental demonstrations came to be discarded, to be replaced by nature study lessons which were believed to lend themselves more readily to heuristic methods. Children in the upper classes of elementary schools began to take elementary science as a class subject, and object lessons were gradually superseded by general elementary science.

As in the previous century, the Board of Education provided useful 'suggestions' each year to help teachers. In due course the *Suggestions for Teachers* moved away from the old disconnected object lessons of a purely descriptive character to a different scheme of teaching. Over the years these 'suggestions' have played a prominent role in shaping thought on science teaching in schools.

Much of the comparatively advanced work was done in the Higher Elementary Schools, and in Central Schools, which came to be built in significant numbers in large towns and cities after 1910. These 'central' schools continued the general education of pupils specially selected at 11 to take advantage of a four-year course having a commercial or industrial bias. As such the curricula of many of these schools had a marked scientific or technical and vocational character. Schemes of general science were introduced, consisting largely of practical work but demanding little theoretical knowledge. Thus boys aged 11–14 tackled the mechanism of electric bells, switchboards, electric motors, air-pumps, hot water supply and so on.

Teacher Training

Lack of an adequate number of trained elementary school teachers with a sound knowledge of science, particularly physical science, was a major obstacle to more effective science teaching in the elementary schools. As the number of children taking science in the elementary schools grew, the teacher training colleges made science compulsory for all their students. In 1907 elementary science was included among the compulsory subjects of the Board's final examination for students in teacher training colleges. In addition, practical work in science was included in the course of study of every two-year student. Every college was obliged to have at least one well-equipped laboratory. But despite the regulations, the work done was of a low standard. In 1908 the inspectors did not report favourably on the attainments of students in the training colleges. One must remember that at this time there was not the same incentive to study science as in the days when the possession of South Kensington (Department of Science and Art) certificates was the high road to a teacher's promotion.

Secondary Education

At a higher (secondary) level there were two types of school:

Division A schools were formed from the old 'science' schools and devoted most of their time to teaching science. They received government grants for teaching specific science subjects. Early inspections conducted by the Board revealed that in many new schools there was almost exclusive attention to grant-earning subjects such as science.
Division B schools devoted only one-third of their time to science, mathematics and manual training.

Thus science formed a part of the curriculum of every secondary school getting grants from the Board of Education.

In 1902 there were 27 852 pupils following organized day courses of instruction in 214 Division A schools. Of the 144 Division B schools, 69 were endowed grammar schools. The smaller endowed schools were not able to establish suitable provisions for science teaching on the government grants alone, and so in these schools science instruction tended to be unsatisfactory.

In the early years of the Board of Education's administration secondary schools enjoyed a generous ration of science. But this amount was gradually reduced. There was a reaction to this science-biased course of study. The evidence of two HMIs focused on the 'undue concentration' on science instruction. In 1902 Buckmaster reported that the timetables were distinctly overloaded in favour of science. Whilst the sciences were generally well taught, the same was not true for English or literary work. J.W. Headlam pointed out that while large grants were available for teaching science subjects pupils were charged extra fees if they wished to study Greek. There was also a lack of facilities for history, literature, etc. in marked contrast to the laboratories and apparatus provided for science.

The 1904 Regulations abolished the distinction between Division A (heavily biased to science) and Division B (less science) schools, abolished grants for individual subjects and introduced a single uniform scale of grant. The abolition of specific subjects and their prescribed syllabuses along with payment of grants diminished the status of science within both the secondary and elementary curriculum. Science no longer enjoyed a privileged position. There was a gradual transformation from curricula predominantly scientific and mathematical to a more balanced curriculum. All secondary schools in receipt of grant were required to provide an approved course of general instruction as a connected and coherent whole over a period of four years. The minimum times for each subject were specified: at least $7^1/_2$ hours a week to teaching science and mathematics and eight hours a week to English, history, geography, and one foreign language. The 13 hours of instruction in scientific and allied subjects which had been compulsory in Division A schools were to be replaced by at least five hours' science teaching and must occupy a minimum of three hours each week for practical work. The practical instruction had to be given in lessons of at least two periods' (1 hour 20 minutes) duration and 'the timetable of the school must be arranged to leave sufficient time to the science teacher for laboratory preparation.' Of the minimum time of three hours a week for practical work not less than two hours must be occupied in practical work by the pupils. For the theoretical instruction not more than 40 pupils were to be taken by one teacher and not more than 20 for practical work or more than 30 if laboratory assistance was provided.

In the higher elementary schools one laboratory for physics and another for chemistry were recommended as desirable. In the schools under the London County Council nature study was taught in the lower standards, whilst in boys' schools simple laboratory work in chemistry and physics was undertaken. Such practical work involved exercises in weighing and measuring and simple manipulations in mechanics, heat, chemistry and electricity. However, a London Technical Education Committee which looked into every aspect of science instruction in the capital reported that, with a few brilliant exceptions, effective practical work was wanting in the schools. Practical work in secondary schools was almost entirely qualitative chemical analysis having no connection with any chemistry taught in the classroom. The Committee reported practical physics was almost entirely absent.

The Public Schools

In spite of the efforts made during the 19th century (see Chapter 9), as late as 1918 the Thomson Committee reported that in the public schools as a whole there has been 'no general recognition of the principle that science should form an essential part of secondary education'. Of the 224 Higher School Certificates awarded by the Oxford and Cambridge Board to Eton, Winchester, Rugby, Cheltenham, Clifton and Marlborough in 1902, only 29 were in science. But this was not so much a reflection of public schools' lack of interest in science as a sign of the discouragement given to science by the pattern of awards at Oxford and Cambridge, by the 'blackmail of the Greek Language specialists'. To gain the maximum number of awards the public schools encouraged their most able pupils to specialize in the classics. Few pupils from the public schools took the preliminary, intermediate or matriculation examinations in science of the University of London. As a correspondent in *Nature* pointed out in 1902, these examinations which were held in high esteem 'scarcely touched the education of the country as represented by the great public schools of England'.

However the report of the Association of Public School Science Masters of 1909 claimed that about 90 per cent of those passing through the public schools in 1907–8 were compelled to take 'a longer or shorter course of science' and that 'laboratories and classrooms specially designed and equipped for science teaching were the rule'. The sort of physics taught in the public schools can be gleaned from the Scheme of Work followed in a boys' public school in 1908–9.

In Forms I and II nature study, including elementary plant physiology, and 'objects in common life' such as air, water, clouds, rain and snow, electric light, flame and the Bunsen burner, the electric bell and some elementary astronomy formed the basis of science teaching. In Form III there was a division into an A group, studying mechanics (including graphs, springs, levers, moments, the balance, steelyard, forces, centre of gravity and 'mechanical principles') and heat (comprising temperature, the thermometer, scales, the measurement of heat, expansion of solids, liquids and gases and hygrometry); and a B group, studying the measurement of length, area and volume, the simple effects of heat; the thermometer, weighing, the balance, simple chemical operations and construction of apparatus, and an examination of air and water. Form IV followed a similar syllabus to that for IIIA, 'rather more fully', and added light (shadows, reflection and refraction of light at plane and curved surfaces, the formation of images, simple optical instruments and the spectrum).

However, by 1918 the majority of public schools provided adequate opportunities for the study of science to those boys whose parents desired it.

Measurement

At the beginning of the century physics teaching consisted largely of accurate measurement and weighing, to engender habits of careful observation. Measurement was seen as the first stage in a systematic study of science. For young children such courses were understandably boring and uninteresting. The balance, the thermometer, the burette, figured largely in laboratory work. For many the first-year course in science was more a course in practical mathematics and frequently killed any interest in science. As Professor J.J. Thomson had reported to the British Association for the Advancement of Science in 1896, 'the student commences his career by knowing how to measure or weigh every physical quantity under the sun but with little desire or enthusiasm to have anything to do with them'.

For this excessive emphasis on measurement Armstrong must take some blame. For Armstrong all true science rested on exact measurement. Speaking at a Conference at Guildford in June 1902 on 'Training in Scientific Method as a Central

Motive in Elementary Schools' he said:

> Measuring appliances come first and foremost: they are indispensable. All true science is based on exact measurement – the introduction of measurement work is the distinguishing feature of modern improvements in our methods of teaching 'science'; the value of 'science' as a school subject is largely due to the fact that it necessitates the introduction of measurement work into the course.

For Armstrong the balance was the ideal piece of equipment.

A British Association report on The Teaching of Science in 1908 recommended that the teaching of mensuration and elementary physical measurement should rather form part of the mathematics teaching. Though quantitative work was important, the qualitative also deserved encouragement.

Expansion of Secondary Education – the Shortage of Science Teachers

In 1910 a new type of school – an Intermediate School – was defined as 'a school providing at least a three years' course of instruction in Language, Mathematics and Science'. In most such schools the time given to science was reduced to three hours a week, the minimum required by the Code, so as to give sufficient time for Language and Mathematics. (In 1936 the time devoted to science was reduced so as to increase the time for PE and music.) This led to much of the practical work being dropped and being replaced by demonstrations.

The expansion of secondary education in the early years of the century was encouraged by a system of 'free places' introduced by the Board in 1907. Under the 1907 Regulations 25 per cent of places in grant-aided secondary schools were to be given to pupils who had spent the previous two years at a public elementary school. But there was difficulty in finding teachers with sufficient training in experimental science. Dr McNair remarked that most of the science masters in the Higher Grade Schools were learning what they had to teach in the process of teaching it. Summer courses were held for teachers to help overcome this difficulty. There was a sharp and sudden rise in the number of applicants for admission to the Royal College of Science in South Kensington in response to the great national shortage of competent science teachers. The new teachers required by the expanding educational system came from two principal sources – the training colleges and the universities. In 1900 there were 61 colleges training 5608 students. By 1913 the number of colleges had increased and double the number of teachers were being trained. The fact that teaching was about the only profession open to university graduates of science meant most of these became schoolmasters, even in the elementary schools, for the opportunities for research in science were limited. Most of the 6000 or so science graduates produced by the universities between 1902 and 1914 went into secondary school-teaching.

Yet even this number was not sufficient to meet the demand for graduate teachers of physics and chemistry in the schools. Whilst 33 per cent of those who took degrees in science had signed a bond (as had training college students) to teach for seven years, the percentage of science graduates who took up teaching was double this number. More and more Day Training Colleges attached to universities and university colleges came to be established. In 1913, in response to pressure from training colleges, the regulations were amended to allow students to take fewer subjects but to study these to a higher level. It became possible for students to obtain their final teaching certificate without having studied any science at all. The result was a rapid falling off in the number of teachers qualified to teach science.

External examinations

External examinations have played – and continue to play – a very important role in science teaching in schools. Much of the development of science teaching in schools and in technical education during the latter half of the 19th century depended entirely on grants made through the examinations of the Science and Art Department. From 1902 onwards the Board of Education Inspectors found that the schools were expressing some dissatisfaction with the examination system. The inspectors themselves felt examinations were leading to over-pressure and the long syllabuses hindered reform of teaching methods. The question of examinations was investigated by a Consultative Committee on Examinations in Secondary Schools which reported in 1911. As a result of its recommendations the Board of Education decided to reorganize the examination system.

In 1914 the Board proposed two examinations:

(a) a **First School Examination** (School Certificate) for pupils aged about 16. The subjects were divided into three main groups: (i) English subjects, (ii) languages, (iii) mathematics and science. Candidates had to reach a reasonable attainment in each of the three groups and in at least five subjects. But since one could pass group (iii) by passing in mathematics without doing science at all, the Board made science compulsory in grant-earning secondary schools. The fact that science was not a compulsory subject for the First School Examination led to science being neglected in other schools.

(b) a **Second School Examination** after two years of specialization in one group. Science was not compulsory except in the mathematics and science group. The Board pointed out that the examination should follow the curriculum, and not determine it.

In 1917 the Secondary Schools Examining Council was set up as a co-ordinating body. This body did not make science compulsory in the First School Examination. Science remained with mathematics in group (iii). This suggested that science and mathematics were alternative subjects. Less than 15 per cent of 800 candidates from the public schools offered science. In *The Aims of Science Teaching* (1917) the Science Masters' Association stated there was no alternative to making science compulsory in the examination: 'if there is no way of making science training in schools both general and adequate except by compulsory examinations in science ...

such compulsion must be employed.'

World War I had impressed upon the nation the vital importance of science as a school subject. The Report of the Prime Minister's Committee which was published in 1918 favoured making science for all candidates compulsory in the First School Examination. In 1919 the Secondary Schools Examination Council again attempted to make science compulsory. But the recommendations were to no avail.

Physics in Girls' Schools

When state-aided secondary schools for girls were set up throughout the country the requirements of the Board of Education necessitated in most cases the teaching of physics and chemistry. But gradually the view became established that these subjects were either too difficult or unsuitable for girls, and if taken at all should be taught with a definite domestic bias. Science lessons were to be intimately correlated with cooking, laundry and other household arts. In girls' schools the usual science subject to be taught came to be botany: no expensive laboratory equipment was required and the subject was 'suitable for ladies' and was supposed by convention to appeal particularly to girls. In course of time, however, the opinion of mistresses in girls' schools turned against such domestic science schemes and the science work slowly came to approximate that of boys.

It was the imbalance in the scientific education of girls established early in the 20th century, with botany the dominant science subject, that led to the under-representation of girls among entrants to the public examinations in physics. The number of physics graduates produced annually between 1925 and 1938 were insufficient to support any expansion of the teaching of physics in girls' secondary schools, especially at sixth form level. As Jenkins points out, in 1962 only 40 per cent of girls' grammar schools in England taught *any* physical science in the fifth year, a figure which falls to nearer 20 per cent when any general science is excluded.

Teaching Methods

In the course of time the heuristic method began to run out of steam. Few teachers actually used the method in their teaching. Heurism was accepted in principle but not in practice, for heuristic teaching is more easily said than done. There were unmistakable signs that in the hands of inexperienced teachers the method was being misapplied; Armstrong's ideals were becoming debased in a heavy emphasis on physical measurement. Increasingly the heuristic method came under attack.

Armstrong had argued his case for heurism in terms of the widely accepted 'faculty' psychology which believed that the mind consisted of several faculties such as 'reasoning', 'imagination', 'memory' etc; and that these could be developed by training and experience. It was also believed that training in one field could be transferred to another field. Throughout the 19th century such a 'faculty' psychology had been used to legitimize the aims, contents and methods of the school curriculum. Heuristic methods 'trained the faculties of thoughtfulness and the power of seeing, producing accuracy of thought, work and of deed'. Armstrong carefully distinguished between scientific fact and scientific method: 'I care little about what is taught, provided the method be sound.'

But the growing influence of the ideas of Froebel, of Pestallozzi, and of Herbart undermined the psychological basis of heurism. These psychologists placed emphasis on freedom rather than on method, on self-development rather than on faculty training. The Herbartians emphasized the role of the teacher in moulding the child. Their stress on the significance of books and literary studies was in conflict with those educationists who valued practical studies, such as Armstrong. The rapid and widespread acceptance of Herbartian ideas, the decline of faculty psychology as a learning theory and the hardening of opinion against 'transfer of training' paved the way for teaching methods based on a different perception of learning. It became increasingly difficult to argue convincingly that pupils must be put in the position of an original discoverer and to maintain that science owed its achievements to a method, 'a game', the rules of which could be learnt and applied. Science could not claim a monopoly of skills associated with activities such as observing, testing hypotheses, thinking logically, verifying provisional concepts. The Herbartian ideas and the decline of faculty psychology paved the way for more consideration of the *subject matter* in the school science curriculum. A Committee of the British Association in 1917 found that the method of science could not be regarded as separate from the matter. This Report on *The Method and Substance of Science Teaching in Secondary School with Particular Reference to the Essential Place of Science in General Education* pointed out that if method were merely the careful observing of facts, their classification, the framing of hypotheses and their verification, then the classics or history master could just as well teach 'scientific method'. As Jenkins points out the new problems became those of matching the subject matter and order of its presentation with the maturing intelligence and developing interests of the pupils.[1]

Appended to the 1917 Report of the British Association for the Advancement of Science were some excellent contributions on the principles of science instruction, the best being by Professor P.C. Nunn, who in 1905 had advanced the notion that the interest of children in science exhibited a rhythm corresponding to the rhythm of history. He identified three motives in man's quest to understand the world around him – 'wonder', 'utility', and 'systematization'. These three motives predominate to different extents at succeeding stages of a child's life. Up to the age of 11 'wonder' was the principal motivating force in holding children's interest in science. Between 12 and 16 'utility' was the main motivating force. It was only at adolescence that the desire for systematization asserted itself. Such ideas, of course, had obvious implications not only for the sequencing of topics within a school science course but also for the method by which they should be taught. Thus, at the earlier age, mere physical measuring should be replaced by studies more likely to appeal to the 'wonder' motive, such as nature study and astronomy. Archimedes' Principle should be taught not as 'a property of fluids or as a method of determining specific gravities but as a principle which explained why ships float' . Between 12 and 16 the utility motive is exercised by examining the working of the electric bell, the dynamo, the wireless set, etc. not as applications of science, but 'as the force of interest from whose study the pupils' knowledge of the scientific principle is to emerge'.

It was understandable that Armstrong should prove hostile to such ideas, with their reduced emphasis on laboratory exercises. 'The "damned boy" needs drilling. We forget this and ever twaddle off playing on his interests.'

World War I highlighted the general lack of scientific knowledge among the ordinary people. Not surprisingly, a letter appeared in *The Times* of 2 February 1916, entitled 'The Neglect of Science', blaming the state of the war on the neglect of science in British education. The letter led to the formation of The Neglect of Science Committee. There was a widespread demand for more science in education. The nation having learned the necessity of science in time of war was ready to admit its value in time of peace. The Government appointed a Committee under the chairmanship of Sir J.J. Thomson, President of the Royal Society, to 'inquire into the position occupied by Natural Science in the Educational System of Great Britain'. The Committee reviewed every phase of science education from elementary education up to university level. It took note of the Science Masters' pamphlet *Science for All*. Its Report – *Natural Science in Education* – was published in 1918.

The Report was able to identify the general pattern of subjects taught in two types of secondary schools, A and B, classified on the age at which the pupils left school – between 15 and 16 in group A, compared to 18 to 19 in Group B. The Committee found that science, though brought down from its exalted position of the Organized Science School days, still occupied 'a position in no way inferior to that of any other subject' in grant-earning schools. Indeed out of 249 scholarships and 96 exhibitions at Cambridge won by boys from 113 grant-earning schools between 1906 and 1915 in all subjects, the number awarded for science was larger than for any other subject. The Thomson Committee attributed this to the stronger position of science and mathematics in sixth-form work compared to classics or modern studies.

The Thomson Committee recommended:[2]

(1) That Science be included in the general course of education for all pupils in Public and other secondary schools up to age 16, 'based on a preliminary course of nature study and the simpler parts of that branch of knowledge which Huxley called physiography (including elementary astronomy), and that this general course be followed by a more systematic study of physics, chemistry and biology'.
(2) That in the First School Examination all pupils be required to satisfy the examiners in both mathematics and in natural science.
(3) That there be co-operation in this Examination between teachers and examiners and that weight be attached to a pupil's school record.
(4) That many more scholarships be available to enable technical students to enter universities and senior technical schools or colleges.

The time accorded to science teaching in all secondary schools was not to be less than four periods a week in the first year of the course from 12 to 16 and not less than six periods a week in each of the succeeding years.

The Report pointed out that the evidence showed there was excessive attention to practical work to the neglect of lectures with illustrative experiments. 'In many schools laboratory practice among junior pupils has degenerated into a routine of quantitative measurements which belong rather to practical Mathematics than to Natural Science.'

While admitting the value of laboratory work by the pupils themselves, the Committee pointed out the dangers of overdoing it. There was room for the demonstration lesson, and the giving of scientific information which the pupils could not easily get was not always out of place. 'Insistence on the view that experiments by the class must always be preferred to demonstration experiments leads to great waste of time,' wrote Thomson. The Committee stated that

> the heuristic method can never be the main method by which the pupil acquires scientific training and knowledge. He cannot expect to rediscover in his school hours all that he may fairly be expected to know; to insist that he should try is to waste his time and his opportunities.

Armstrong was furious at this caricature of what he had never advocated. For Armstrong heurism meant 'directed inquiry' with the teacher playing a positive but discreet role.

Though science formed a regular part of the elementary school curriculum, the standard of attainment was low. The Committee held that this was due to the low qualifications of the teachers. 'It is extremely desirable that there should be a much larger number of teachers in Elementary Schools qualified to give instruction in Science and all possible steps should be taken to increase the supply.' The Committee suggested this should be done by special classes and short courses for teachers. It strongly deprecated the use of peripatetic teachers.

For the 16–18s the Committee suggested that time be found for lectures on the history of science as exemplified in the lives of the great scientists. There should be some elementary treatment of the development of the leading concepts of science. Science was to be regarded as another of the humanities – another great highway of human endeavours.

The Rise of the General Science Movement

The most important change that took place in school science teaching in the first half of the century was the development of general science. By the beginning of World War II it had occupied a position on a par with that of the separate science subjects. Before the turn of the century the University of London had substituted the subject general science for the specific sciences in its matriculation examination.

During the second decade of the century the Board of Education's *Suggestions for Teachers* came to advocate a broader spectrum of science in the elementary schools, away from arid exercises in measurement and in specific-gravity determinations and in verification of Boyle's Law. The 1915 *Suggestions* proposed a general course of lessons centring round the properties of air and water, or around the mechanics of simple machines. Such courses could comprise a series of problems each occupying several lessons involving oral work, demonstrations and some individual practical work. For older boys the Board suggested that a scheme of work in physics might include

(i) the solution of practical problems such as the fitting up of an electric light inside a cupboard so that the opening of the door turns on a light and the shutting turns it off, or the devising of a water motor, the making of a small windmill to work a pump etc.;
(ii) the making of apparatus such as a model telegraph or telephone or garden pump;
(iii) the repairing of apparatus such as an electric bell or a defective lock.

In 1915 a few teachers in the Association of Public School Science Masters (founded in 1906) started a 'Science for All' campaign which aimed at a more general course of science teaching. The Consultative Committee on Examinations in Secondary Schools had pointed out in 1911 that it would appear that 'in the upper forms of many schools the course of work is largely modelled on the needs of the few pupils who intend to proceed to the University, the school studies of the majority being relegated by the special requirements of the few.'[3]

In 1916 the Association proposed a course of general science in response to growing demands for a wider knowledge of science. The idea behind the general science movement was the elimination of 'academic' conventional teaching from the science course for pupils up to 16. A broader science should be part of the general education of boys and girls up to that age. Practical work throughout the course was regarded as essential, 'as much as possible being done by the boys' and 'illustrated lectures' for some topics. The scheme was intended for the middle forms of secondary schools. General science was to be a sequel to introductory courses of science and a prelude to later specialization.

The subject of general science received a fillip in 1921 when it was approved as a Group 3 subject by the Oxford and Cambridge Joint Board. Candidates were required to sit two papers, each of two hours, which included questions on physics, chemistry, biology, astronomy, geology and meteorology. However, many of the questions set were more appropriate to separate papers in physics and chemistry. That specialization in science subjects in the secondary schools continued unabated is evidenced by the fact that in 1925 only two schools out of 39 in England presented boys for the general science paper of the Oxford and Cambridge Joint Board.

In a Presidential address to the British Association at its Hull meeting in 1922 Sir Richard Gregory pointed out that science teaching should not be to prepare pupils for vocations but to equip them for life; science in schools must not be planned in the assumption that pupils will proceed to a university course. Less than three per cent of the pupils of state-aided secondary schools went to university and yet most of the science courses in these schools were based on syllabuses geared to university entrance examinations, rather than on what should be placed before young and receptive minds as the scientific portion of their general education. He advocated 'Science for all'.

There were many reasons for the continuation of specialization in separate science subjects. There was the entrenched positions of physics and chemistry as separate subjects in the curriculum. Also, the headmasters of secondary schools preferred teachers with specialist honours degrees and so trainee graduate science teachers would opt for separate subjects. Finally, the general science movement fared less well because the hybrid subject lacked coherence and was widely taught as a scrappy mass of isolated facts from the different sciences.

Teaching general science is difficult. It is not just a case of providing information gleaned from several fields. The essence of general science lies not in content but in interpretation and that involves teachers of wide experience, and outlook. The Science Masters pointed this out clearly in their *General Science* (1924). The teaching of general science makes greater demands on the teacher than physics or chemistry taught along conventional lines. On paper a syllabus of general science is just a list of topics covering a very wide field. It is only a very skilful and experienced teacher who can provide the unity required.

The Board of Education report 1924–5 points to improvements in the teaching of physics in schools.

> The teaching of science is now completely revolutionized; more qualitative work is done, whereas previously practically all the work in physics was quantitative, the scope was now much broader, more branches of physics were studied, more interest was created among pupils, and as Mr Frewin states: 'the numerical conundrums that inevitably followed experiments, as if science had been brought into schools to provide practical material for Arithmetic, have had their day'.[4]

Whilst the Thomson Committee had gone to some lengths to condemn the worst excesses of heurism in schools it favoured no one teaching method. The Hadow Report of 1926 gave more attention to content than to the methods of school science teaching. A general science syllabus was suggested for Modern Schools or the senior classes of ordinary elementary schools which was grouped round four main divisions:

(1) Elementary chemistry and astronomy and meteorology;
(2) A carefully graduated course in elementary physics and simple mechanics, abundantly illustrated by means of easy experiments in light, heat, sound and the various methods for the production and application of electricity;
(3) Biology;
(4) Elementary physiology and hygiene.

The syllabus added: 'In order, however, to bring home to the children the practical application of science to everyday life, the lessons should be fully illustrated at every possible point by reference to the environment.'[5]

The Hadow Committee felt it was desirable to have a special leaving examination for Modern Schools and Senior Schools, not only as a stimulus for pupils but because otherwise these schools 'would aim, as some of them do now, at presenting considerable numbers of pupils for academic examinations such as the First School Examination and the London Matriculation'. The First School Examination was considered inappropriate because

(i) it presupposes a course of teaching extending over four to five years, whereas in Modern Schools and Senior Classes the course will, as a rule, be limited to three or four years.
(ii) in its present form the examination appears to be either too literary or restricted in scope.

Practical examinations

A British Association Report in 1928 – *Science in the School Certificate Examination* – criticized the lack of a practical test in the School Certificate examination.

> There is no practical examination and no safeguard against cramming either in the syllabus or in the nature of the questions. It is scarcely too much to say that a candidate could pass the examination without possessing any real scientific knowledge of scientific principles or experimental methods of study. The interpretation of the syllabus by the examiners, as shown in the questions set, often shows a misconception of what general elementary science or the science of everyday life should signify.

The Report of the investigators into Group 3 (i.e. science subjects) of the First School Certificate recommended that a practical test be held for such subjects in the examination. 'The main object at this stage should be to supply evidence that a Candidate has done laboratory work and had been taught on proper lines, and it is important that wherever possible a Practical Examination should be held.'

Teachers were divided on the question of practical examinations. Whilst all agreed practical work was essential, some felt practical examinations had a cramping and limiting effect. The tasks set often merely tested certain stereotyped exercises which involved no real experimental work in the true sense. By 1928 a body of opinion had grown up which questioned the desirability of holding practical examinations. Many teachers urged the abolition of practical examinations at the First School Certificate stage. At the second stage it was felt practical examinations were necessary to ensure candidates had acquired mastery of certain techniques.

The investigations in 1931 reported that 'the need for compulsion (of practical examinations) had passed'. They stated that there was no difference between schools which had a practical test and those which had none in the importance they attached to practical work and the thoroughness with which it was done. They suggested that laboratory work and experience should be tested by questions incorporated in the written examination.

By 1930 six of eight Examining Boards for the School Certificate were offering a syllabus and an examination in general science. By 1931 general science was making some inroads into the curriculum of the grammar schools, though physics and chemistry continued to dominate the science work in boys' schools. General science was usually taken as an introductory course for pupils aged 11 to 13.

In 1936 appeared the Science Masters' Association influential *The Teaching of General Science Part 1*. The syllabus recommended for general science was somewhat anomalously printed in three columns headed *physics*, *chemistry*, *biology*. The topic method was suggested for teaching general science. A topic – air, water, radiation, energy, etc. – would be chosen and then studied from several aspects. A topic on energy from the Sun could for example, be developed along the lines shown in Fig.16.1. Advocates of the topic method argued that it allowed teachers to bridge the conventional boundaries between physics, chemistry and biology. The purely didactic method of teaching was dismissed. The Committee favoured the **problem method**. The teacher begins with some ordinary phenomenon or problem.

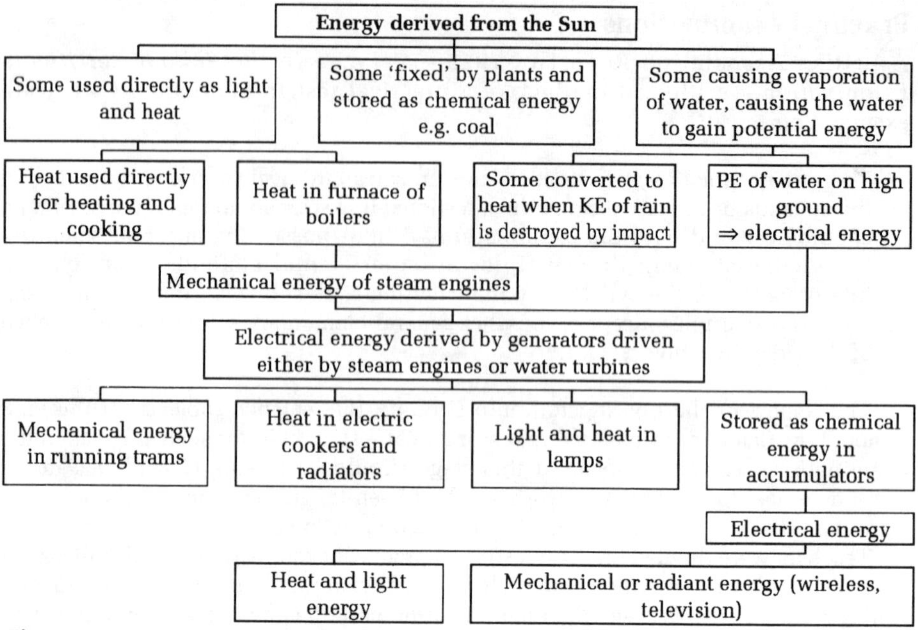

Figure 16.1

Discussion leads to a clear statement of a problem which can be answered by careful experiment, and suggestions are obtained from the class as to how the apparatus is to be arranged to obtain an answer. Finally, the results are gathered together and from them a law or general statement emerges.

In the senior schools only slow progress was made in broadening the science teaching during the 1930s. The Board of Education's Handbook of *Suggestions for Teachers* had admitted that successful science teaching in the senior schools often required teachers to abandon the 'academic' approach in which they had been brought up. Their failure to do so contributed largely to the lack of progress of general science in the senior schools. However, towards the end of the 1930s the need for a broader school science curriculum came to be accepted more and more. General science as a subject increased after 1939, as the figures in Table 16.1 show.

Table 16.1 **General Science First School Examination**

	1919	1924	1929	1934	1937	1938	1939	1942
Number of entries	518	1266	1501	2842	4779	8784	12 497	7817
Percentage of candidates taking the subject	1.8	2.6	2.5	4.3	6.2	11.4	15	23

This was probably due to the exigencies of the war and to the difficulty of obtaining specialist science staff. Whereas in 1937 the percentage of all candidates offering general science as a subject in the School Certificate was only 6.2 per cent, by 1943 it had risen to 23 per cent. General science had achieved a position as an examination subject close to that of the three main science subjects. The number of schools entering candidates for general science rose from 240 in 1936 to 680 in 1942. By 1943 about 40 per cent of secondary schools in England and Wales presented candidates in general science for the School Certificate. But where general science was one of a number of science subjects offered in a school, it tended to be taught to weaker candidates and non-science specialists, usually as a course of three or four years' duration up to School Certificate.

After the war the position achieved by general science in relation to single science subjects during the war was not maintained. There was a steady, proportional decline in the number of O-level entries for general science during the 1960s and by the mid-70s most GCE examining boards had withdrawn general science as an examination subject.

There were several reasons for the decline in the general science movement. One was the persistent criticism that general science was just a miscellany of odds and ends lacking any coherence or sound integrating principles. Even the general science syllabus produced by the Science Masters' Association presented the contents of the course under the conventional headings of physics, chemistry, biology. Furthermore, as Jenkins points out, the more radical aspects of general science were in conflict with the pre-professional function ascribed to grammar school courses in science subjects, a function which was strongly supported by the structure and format of the public examinations.[6]

Post-war society was committed to the production of highly qualified scientists in greatly increased numbers. Also, within the schools there was a massive increase in the number of sixth form pupils. Between 1955 and 1965 the number of pupils in the sixth forms doubled. During the 1950s there was an amazing swing to the science subjects in the sixth forms. By 1960 more than 60 per cent of sixth form boys were specializing wholly in science and/or mathematics. These developments reinforced the pre-professional function of the grammar school science course. In such a climate general science could not hope to compete.

In the secondary modern schools general science also declined, handicapped by the shortage of qualified science teachers and the suspicions entertained by employers about the value of general science.

The Supply of Teachers

In contrast to the 19th century, the period between the two World Wars was one of a chronic over-supply of teachers. In the 1930s the number of science posts available in secondary schools was about 300, but there would be 1000 newly qualified teachers seeking these jobs. Headmasters would receive 100 applications for one post in a grammar school. In order to get a job some science graduates were taking posts to teach PE or woodwork. College-trained elementary and secondary school teachers faced greater difficulties. At the end of 1932 about 1100 newly qualified teachers remained unemployed. Admission to training colleges was cut by $2^1/_2$ per cent. Further cuts of 10 per cent and 8 per cent followed in 1933 and 1934.

After the war there was a chronic shortage of science teachers: industry had replaced teaching as an occupation of those with science degrees. An Emergency Teacher Training Scheme was introduced at the end of the war. In 1947 the school-leaving age was raised to 15. By 1957, when the grammar schools were accommodating greatly increased numbers of pupils in their sixth forms, at least 250 science posts remained unfilled, 222 were filled unsatisfactorily and 614 posts were held by non-graduates. The secondary modern and girls' schools probably suffered the most from the shortage of science teachers.

The Nuffield Science Teaching Project

By 1937 the heuristic approach to science teaching had virtually disappeared, killed by the examination system, the criticism of the new psychology, the general science movement and by brilliant writers like F.W. Westaway, who in 1929 wrote an excellent book *Science Teaching* which, whilst recognizing Armstrong's efforts to improve science teaching, was explicit in rejecting heurism as too slow and unworkable. The effectiveness of the heuristic method of teaching depends on the ability of the teacher. And it was the teachers who were the method's weakest point.

To the end Armstrong remained convinced of the importance of training pupils in scientific method; and despite the virtual eclipse of the heuristic method of teaching science, many of his ideas continued to influence school science education. An emphasis on practical, experimental teaching and a belief in the importance of learning by doing became recognized features, and a generation after Armstrong's death in 1937 curriculum reform in science, as evidenced by the Nuffield Project in Britain and the Physical Sciences Study Committee (PSSC) in the USA, was to incorporate his view that science could best contribute to a liberal education by initiating pupils into its greatest professional mystery, its method. The Nuffield Science Teaching Project stressed the teaching of science 'as an inquiry' and encouraged pupils 'to think about scientific things in a way that practising scientists do'. In so far as the Nuffield Science Teaching Project placed particular emphasis on the *process* of scientific investigation., on discovery learning, it could claim to be a recognizable ancestor of Armstrong's heurism.

It was in the 1950s that the first stirrings of present-day curriculum reform began to manifest themselves. These took concrete shape in the 1960s with numerous projects all of which advocated discovery ('inquiry') methods of learning. The view of science which informed the Nuffield School Science Projects in Britain and projects such as the PSSC in the United States was 'science as search'. By emphasizing a discovery approach it was hoped that both the content and method of science could be learnt simultaneously. John Lewis of Malvern College gives a very revealing account of the transition from the more 'traditional' to the more 'modern' approach to physics teaching (see Appendix 10). He describes his early years of physics – the pin-sticking process for finding refractive indices, finding the focal lengths of lenses, plotting magnetic fields, confirming Hooke's Law and Ohm's Law, exercises on the Wheatstone bridge. Higher up the school he worked with thermionic valves, and plotted the characteristic curve of diodes. Much of this has now been replaced by electronics.

The Nuffield approach has had a profound influence on all school physics teaching. It has brought out the essential unity of the subject, continually stressing the inter-relationships of the parts, with concepts of energy and ideas of wave motion pervading the whole fabric of the subject. The old compartmentalization of heat, light, sound, electricity and magnetism has gone. Practical work concentrates on developing an understanding of the unifying concepts, rather than on measurements so typical of the past. Using the electric circuit board, young pupils become 'scientists for the day' and find out for themselves the behaviour of electric circuits and the laws which govern them. The pupils learn to think, to search for patterns, to do 'real' physics, instead of learning formulae by heart.

Whilst the Nuffield approach certainly stimulates an enquiring mind, it has one defect in that, as John Lewis points out, it does not sufficiently emphasize the *social relevance* of the subject.[7] But this defect has been remedied in physics teaching since the introduction of the Nuffield approach.

Epilogue

It is not easy to write 'historically' of changes in physics teaching occurring in our own lifetimes. One may, nonetheless, attempt to encapsulate the main features of the last few decades.

If the 1960s was the decade for 'science for the enquiring mind' projects, of which the Nuffield Project was typical, then the 1970s was typically the decade for 'integrated science programmes'. The 1980s took up the issue highlighted by John Lewis (see page 202), namely the missing thread of the inter-relationship between **science** and **society and the environment**. These issues form an integrating thread of the National Science Curriculum for the 1990s.

Appendices

1 Examples of Science Questions set to Students at the End of their Training College Course[1]

1856 Mechanics (Second Year)

1. Define friction and give instances of its uses.
2. Define the centre of gravity and show how the centre of gravity of an irregular body may be found experimentally.
3. Find the centre of gravity of six bodies weighing 1, 2, 3, 4, 5, 6 lbs respectively and placed in consecutive order at the six angles of a hexagon each of whose sides measures 1 foot.
4. State precisely how you would demonstrate to your first class:
   i) the principle of the lever;
   ii) the parallelogram of forces;
   iii) the laws of friction.
5. Write full notes, state the apparatus if any you would exhibit, draw the diagrams you would put on the blackboard for a lesson to your first class on the fire engine.

Physical Science (Third Year)

Section 1

1. Describe the construction and explain the action of
   i) the air pump;
   ii) the differential thermometer;
   iii) the galvanic battery.

Section 2

1. Describe the parallel motion (sic) and demonstrate its principles mathematically, or show by diagrams how it could be demonstrated experimentally. What different forms may it resume?
3. Explain the construction and use of a sextant.

Section 3

1. Describe the hygrometer and explain its action.

2 University of London Matriculation Examination Syllabus for Natural Philosophy (Pass Examination) 1838

Natural Philosophy*

Mechanics: Explain the Composition and Resolution of Statical Forces; describe the Simple Machine (Mechanical Powers), and state the Ratio of the power to the Weight in each; define the Centre of Gravity; give the General Laws of Motion, and describe the chief experiments by which they may be illustrated; state the Law of the Motion of Falling Bodies.

Hydrostatics, Hydraulics and Pneumatics: Explain the Pressure of Liquids and Gases, its equal diffusion, and variation with the depth; define Specific Gravity, and show how the specific gravity of bodies may be ascertained; describe and explain the Barometer, the Siphon, the Common Pump and Forcing Pump, and the Air-Pump.

Acoustics: Describe the nature of Sound.

Optics: State the Laws of Reflection and Refractrion; explain the formation of images by Simple Lenses.

Chemistry: Heat – its sources. Expansion. Thermometers – relations between different Scales in common use. Difference between Temperature and Quantity of heat. Specific and latent heat. Calorimeters. Liquifaction. Ebullition. Evaporation. Conduction. Convection. Radiation.

* A popular knowledge only of these subjects ... will be required, such as may be obtained by attending a course of experimental lectures.

Natural Philosophy Paper set for the Honours Matriculation Examination, 1839

1. Calculate the force necessary to draw a carriage-wheel over an obstacle on a level road.
2. A plane is supported horizontally by three props: if a weight is placed on the plane, find the relative pressures sustained by each prop.
3. A conical vessel is suspended by two equal strings attached to two points at a given distance in the same horizontal line and connected with the rim of the base at points diametrically opposite; find the tension of the strings.
4. Find the centre of gravity of the sector of a circle.
5. Required the velocity with which a projectile must be cast in a given direction in order to have a given range measured on an inclined plane.
6. P and Q are weights connected by a string passing over a fixed pulley: P descends from rest; find its velocity after t seconds.
7. Find the tension of the rod of a pendulum in vibration when its position is exactly vertical.
8. Calculate the total pressure on the sides and base of a cubical vessel containing a heavy uniform fluid.
9. Give some explanation of the cause of the elevation or depression of fluids in capillary tubes.
10. Suppose two sides of an equilateral triangle are reflectors, find the total number of images which they would produce when an object is placed in the middle of the base of the triangle.
11. Trace the path of a ray of light incident on a spherical drop of water.

Hydrostatics and Optics

1. The surface of any heavy fluid at rest is in a horizontal plane.
2. The surface of a fluid contained in a vessel which uniformly revolves round a vertical axis, will acquire the form of a paraboloid of revolution.
3. The resultant pressure on a solid immersed in a fluid is vertical passing through the centre of gravity of the fluid displaced, and is equal to the weight of that displaced fluid.
4. Find all the possible positions of equilibrium of an equilateral triangle immersed in a fluid, of which the specific gravity is n times that of the triangle.

5. Define the metacentre of a floating body, and show how far the stability of the equilibrium of such a body is dependent on the relative positions of the metacentre and centre of gravity.
6. Describe some experiments for ascertaining the velocity of water issuing from a small orifice in the side or bottom of the containing vessel, both at the orifice and at the vena contracta.
7. Show that the intensity of light emanating from a surface is proportional to the line of the angle of emanation; what effect is produced by this law on the apparent brightness of the different parts of the sun's disc?
8. The image of a body produced by a plane reflector is exactly equal in dimensions and similar in form to the object. Account also for the two images produced by a common looking-glass when viewed obliquely, and for the extreme faintness of the second image.
9. Find the relative positions of the conjugate foci when diverging or converging rays fall on a polished spherical surface, measuring from this surface.
10. Find the deviation of a ray of light falling at a given incidence on a glass prism.
11. Give an account of the methods which have been adopted to produce achromatism in systems of lenses; is the eye such a combination?
12. Explain clearly the formation of the primary rainbow and of lunar haloes.

3 Syllabus of Instruction in Experimental Science for Higher Grade (Science) Schools 1898

First year

Introduction. Classification. Differences of air, water and solids.
Terms – solid, liquid, gass.
Measurements of lengths, areas, volumes. Units.
Laboratory practice in the use of the yard-stick, metre-stick, measuring jar and burette. Relation between the British and Metric systems.

> Exercises in measurement of lengths, areas and volumes should not be unduly prolonged; they should cease as soon as pupils can determine with fair accuracy areas of triangles, quadrilaterals and circles in sq. cm. and sq. in. of the laboratory tables or classroom walls or floor in sq. in. and sq. ft.; the volumes of a cube, a square prism, a cone and a cylinder in cu. cm. and cu. ins., the capacity of a flask in c.c. and of the classroom in cu. m. and cu. yds. These measurements are merely a means to an end, but interest will be added to them and their usefulness increased if employed to find out or to verify certain well-known relations, e.g., the circumference of a circle and its diameter.

Verification of well-known proportions.
Area measurements to establish some of Euclid's propositions.
Volume measurements to discover the relation between the volumes of a cone and a cylinder on the same base and attitude.

Relative Densities

The use of the balance will follow. The class now able to find weight and volume is ready for the conception of density. The relative densities of solids should be found, also the weight of 1 c.c. of water, the volume of a small flask (by weighing the water it will hold, and then the relative densities of methylated spirits, milk and sea water with the help of the flask).

The Principle of Archimedes.

Expansion. Solids, liquids and air will be submitted to experiment.

Thermometers. C and F thermometers side by side in a mixture of ice and water heated over a slow flame. The temperature is recorded every 30 seconds till the water boils. A curve should be plotted.

Barometer. Mercury barometer. Hare's apparatus. Application of the principle in the working of pumps, and in the use of the siphon.

Evaporation. Solution and distillation.

Second Year

Chemistry only.

Third Year

A. For boys 'detailed and systematic study in the classroom and laboratory of Charles' Law, Boyle's Law, latent heat, specific heat, melting points, boiling points and the following treated simply: – force; gravitational, magnetic, electric and refraction of light.
The heating, chemical and magnetic effects of an electric current.

B. For girls: – domestic science and biology.

C. For rural schools – farm physiology and agriculture.

4 Physics textbooks in the Public and Endowed Grammar Schools in the second half of the 19th century

It is ironic that the two most popular physics textbooks used in the Public and Endowed Grammar Schools of England during the second half of the 19th century were French textbooks – or, rather, translations of French textbooks. This possibly reflects the greater attention given to science in French schools than in their counterparts in England. The two books in question were excellent textbooks. The content they covered was up to date for the time, and was presented in good, easy-to-understand language, supplemented by diagrams and illustrations, lavish in their number and excellent in high-lighting the facts they were designed to high-light. The two textbooks were: Adolphe Ganot's *Elementary Course in Physics* and A. Privat Deschanel's *Elementary Treatise on Natural Philosophy*.

Adolphe Ganot

Ganot's textbooks were widely used and was used the physics textbook in the following Public Schools: Rugby, Eton, Winchester, Marlborough, Cheltenham, Taunton, King's College School, Wellington, City of London School and Manchester Grammar School.

Privat Deschanel

The other popular (French) textbook of physics was Privat Deschanel's *Elementary Treatise on Natural Philosophy* (1868). This was a translation of his *Traite de Physique* made by Professor J.D. Everett, Professor of Natural Philosophy at Queen's College, Belfast. Deschanel's *Traite de Physique* was adopted by the French Ministry of Instruction as the official physics textbook for Government Schools. It covered the fields of Dynamics, Heat, Electrostatics, Magnetism, Current Electricity, Acoustics and Optics. In very many instances the book goes well beyond GCSE standard, covering ground that is up to today's GCE 'A' Level standard.

Deschanel's textbooks were perhaps even more popular than Ganot's in the English Public Schools of the last half of the 19th century. Deschanel's *Natural Philosophy* was used in Rugby, Harrow, Marlborough, Dulwich, Cheltenham, Christ's Hospital, Manchester Grammar School and Clifton College. Deschanel's *Electricity and Magnetism* was also used in Wellington.

English textbooks

One textbook that could perhaps match the two French physics textbooks was **Dionysius Lardner's** *Natural Philosophy for Schools*, which appeared in 1857, costing 3s 6d. Lardner had been Professor of Natural Philosophy and Astronomy at University College, London. Like the French texts this book was an *excellent* class-book, lavishly illustrated with 328 very good, clear, diagrams and illustrations.

The other English physics textbooks that could come anywhere near competing with the French texts were those by **Balfour Stewart**, Professor of Natural Philosophy at Queens College, Manchester. His *Elementary Physics Primer* (junior) was used at Marlborough, Cheltenham, Taunton, Christ's Hospital, Manchester Grammar School, Clifton and Stonyhurst. His *Elementary Physics Primer* (senior) was in use at Dulwich, Christ's Hospital and the City of London School. Another good physics textbook, again very well illustrated, was Balfour Stewart's Lessons in Elementary Physics (370 pages, 1870). This was a more advanced text, covering up to GCSE standard and even covering 'A' level work in certain areas. Balfour Stewart's Textbook on Physics was used in Rugby, Eton, Winchester and Clifton.

In association with W.W.H. Gee, who was a demonstrator and assistant lecturer in Physics at Owens College, Manchester, where Stewart had been Professor of Physics, Stewart produced an excellent and deservedly very popular textbook called *Practical Physics for Schools and the Junior Students of Colleges* (1888). In the Preface Stewart touches on the main reasons why practical instruction in physics in schools was a somewhat rare phenomenon – lack of trained teachers, lack of any recognized pedagogy for practical instruction in science, and lack of suitable apparatus. It was with a view to remedying this lack of practical work in schools that Stewart and Gee produced their text.

A very popular textbook on chemistry was George Fownes' *Manual of Elementary Chemistry*. Half the book deals with subjects today considered more appropriate to physics, namely heat, light, electricity and magnetism. The textbook was used at Rugby.

5 Specimen Examination Questions set at Eton 1869/1870

These questions were taken from the Appendix to the Sixth Report of the Royal Commission on Scientific Instruction, 1876, pp.125–6.

Machinery. C. (Form V) 1870

1. Show how a movable pulley doubles the force of a workman.
2 Give figures of movable pulleys, hanging by the same, and by different strings. Show what forces balance each other on each of them.
3. Mention different kinds of levers, and state where the force is applied, and where the resistance is overcome in each.
4. At what distances from the fixed point will weights of 10 lbs and 2 lbs balance each other on a straight lever?
5. Draw figures of a wheel and axle, showing clearly how the ropes work, and the advantage of using the machine.
6. What substitutes are used in a windlass and a capstan for a wheel and rope?
7. What substitutes are used in a rack and pinion for an axle and rope?
8. Show how a crane works, and how it multiplies force.
9 When force is multiplied by any machine, show that longer time is spent.

Hydrostatics. B. (Collections East, 1870)

1. Mention some illustrations of the principle that the free surface of a liquid is a horizontal plane.
2. Describe Bramah's press and the contrivance for making it water-tight. Account for the pressure produced by it.
3. Give figures of a common pump, and of an air-pump.
4. Show how to find the specific gravity –
   1. Of a solid; 2. Of a liquid.
5. Explain how it is that water can be lifted in a thin straight open tube.
6. Describe Boyle's experiments with compressed air. What conclusions did he draw from them? How far has a diving bell descended when the air is compressed into half its original volume?
7. Account for the action of a siphon, and of an intermittent spring.
8. Describe the hydraulic ram.
9. What keeps the mercury in a barometer from sinking down into the cistern? If a small hole were bored in the middle of the tube, what would happen?
   How can a barometer give an indication of the probable state of the weather?
10. Describe a single-action steam engine. Explain what forces cause the alternate motion of the beam.
    What is the object –
    1. Of the fly-wheel; 2. Of the governor?
11. What is the use of the small hole in the lid of a tea-pot?

6 Professor G. Carey Foster, On the Teaching of Physics

At the outset Foster argues that it is the scientific habit of mind, rather than mere facts, that is the desired aim of physics teaching.

> A scientific habit of mind is of far more value than a knowledge of scientific facts; in proof of which it is sufficient to call to mind that ascertained facts can nearly always be found recorded in books of reference, while only a trained intellect can interpret the lessons conveyed by any new combination of them.

Nevertheless, facts are important.

> The first object of the teacher must be to impart clear physical ideas, which can only be done by making the pupils get at any rate some part of their knowledge of physical facts through their own personal observation of phenomena....In fact, personal observation and experience naturally produce living ideas, whereas our attempts to communicate knowledge by mere statements and descriptions often result in a form of words being committed to memory quite apart from the acquisition of any corresponding ideas by the mind.
>
> ... Our object should be, not to try to convince [pupils] of the truth of physical principles, but to help them to convince themselves by reasoning upon the facts they have themselves observed. Hence, in choosing practically the particular subjects to be included in the course of instruction, the two following considerations must be kept in mind: first, the greater or less extent to which the fundamental facts of the several departments of Physics can be demonstrated by such experiments as can be performed by the pupils themselves; and secondly; the extent to which the pupils are able to follow the processes of mathematical reasoning needed for the discussion of the experimental results.
>
> Probably the first question which a schoolmaster, who wishes to introduce the teaching of Physics into his school, asks himself, is, in most cases, how much and what kind of apparatus will be wanted? And this question, serious enough when it is only intended that the master should show experiments to the boys, becomes much more important, if the method I have advocated, of letting the boys make most of the experiments for themselves, is to be adopted.
>
> If a whole class of boys are to perform the same experiments at the same time, it is of course necessary to have a sufficient stock of apparatus to be able to give to every two, or to every three boys at most, the things needed for making each experiment. This, however, it would probably, in most cases, be impossible to do, or would be possible only by restricting within very narrow limits the range of work given to the boys. On the other hand, by adopting the system practised, I believe, in some of the American laboratories, the necessity for multiplying very greatly apparatus of the same kind is to a great extent got rid of. Upon this plan the boys working at the same time are engaged upon different experiments, and, therefore, use different apparatus. Previous to each lesson each working place is supplied with the apparatus wanted for some particular experiment, and the boys, as they come into the room, are told which place they are to go to for that

lesson, the place of each batch of two or three boys being changed from lesson to lesson, till all the boys have made all the experiments provided... The advantage of making the boys circulate instead of the apparatus is obvious.

It is clear that, by this method of working, a given outlay on apparatus will suffice for a much greater variety of experiments than it would provide for upon the first mentioned method, according to which all the pupils simultaneously make the same experiment. On the other hand, it greatly increases the labour of teaching a given number of boys, for it excludes the possibility of collective class instruction in the details of each experiment. To make it practicable at all, when one master has to deal with any considerable number of pupils, it would probably be needful to give to each group minute written or printed instructions as to the precautions and manipulatory details required for making the experiments assigned to them. Broadly stated, the difference between these two methods of instruction may be said to be, that the first involves the greater expenditure on apparatus, and the second on teaching power. The choice between them must be decided according to the special circumstances of individual schools.

... The expensive contrivances of the instrument maker may generally be replaced by less elegant but equally serviceable substitutes put together from such things as are usually accessible. And this leads me to remark parenthetically, that a master who wishes to teach experimental physics with success must not be above learning how to solder, to use a file, to drill holes and cut screws, and to use joiners' tools with such skill as he can acquire ...

Even more important than a judicious choice of apparatus is that of the experiments to be made with them when they are bought. With regard to this point, my own experience leads me to think that merely qualitative experiments are of little use in practical instruction; they do not afford sufficient occupation for either hands or head, and are, therefore, apt to degenerate into play. Simple physical measurements, on the contrary, give plenty to do, and yield far more instruction. For instance, if a boy is told to form an image of a flame upon a screen by means of a convex lens, he will have finished the experiment in less than two minutes and want to know what he is to do next; but if he be told to measure the distances between flame and lens, and between lens and screen, at which sharp images can be formed, for half a dozen prescribed positions of the screen, and also to measure the size of the image in each case, he has got work enough to keep him attentively occupied for a full hour, and at the end of it he has acquired data from which, with a little guidance, he cannot help deducing the principal laws of the action of lenses.

... But in order to make sure that the pupils know clearly what they are intended to do, and also to train them in methodical habits of working, it will probably be found useful to give them, with such exercise that is set them, a blank scheme or schedule, to be filled up with the results of their measurements and with whatever further results it is intended that they should deduce by calculation from these.

... For boys who have gone carefully and systematically through even one or two sets of experiments, are far better able than they were before to understand and profit by experiments which they see performed by their teacher, or of which they merely read or hear descriptions ... The relation of teacher to his class should resemble that of a guide, who knows better than the rest of a part the ground that all are to traverse – not that of a traveller, who comes back to show to an admiring audience the wonderful things he has brought from places that they have never seen.

... I still venture to ask you attention to a few points, the importance of which – self-evident as it may seem – was not so fully obvious to myself when I first tried to teach Physics as it has now become.

(1) The most essential of all the conditions for successful teaching is that the teacher himself should have an absolute clear idea of what is to be taught. It is far from being enough for him to have 'got up' the subject carefully from books, and to know exactly what they say about it; he ought to have such a thorough understanding of it as will make him independent of the manner in which it has been presented by others and will enable him to place his pupils at whatever point of view shows most plainly the connexion between the new subject and their previous knowledge ...

(2) In teaching a subject which depends so completely as Physics upon experimental demonstrations, it is needful to be on our guard against letting experiments occupy the place of the principles they are employed to establish, and letting the mechanical details of apparatus obscure the perception of the experiments that are made with them. The kind of fault I refer to is far too common in most of the elementary text-books. For instance, in treatises on electricity, we almost always find one or more chapters devoted to the electrical condenser and Leyden jar where the construction and action of these apparatus are elaborately described, but in which very often there is no clear statement of the nature of electrical capacity, and of the conditions which determine its amount in any special case.

(3) ... In the case of Physics, the frequent use of technical terms cannot be avoided; but we ought always to remember that the use of the study of Physics is not to explain such terms, but that the use of the terms is to facilitate the study. In beginning each new subject, we ought to keep completely out of sight all the technical terms relating to it, and should explain in common language, even at the risk of considerable circumlocution, the ideas that these words are intended to express. For instance, we should not begin by talking about specific heat, or electromotive force, or indices of refraction, and then proceed to explain what the term means; but we should first demonstrate the existence of the property in question, and then, after the idea has been grasped, we should give a name to it. In this way, when the word follows the idea, it is understood at once and felt as a help by the learner, instead of being regarded, as it often is when the opposite order is adopted, as something even more difficult to understand than the idea which it ought to convey.

7 Regulations for the London BA Pass Degree 1839

No candidate shall be admitted within two academic years of passing the matriculation examination. The fee for the examination is £10. Candidates will be examined in Mathematics and Natural Philosophy, Chemistry, Animal Physiology, Vegetable Physiology and Structural Botany, Classics, Logic and Moral Philosophy.

Mechanics: The Composition and Resolution of Forces; the Mechanical Powers; the Centre of Gravity; the general laws of Motion; the Motion of Falling Bodies in free space and down inclined planes.
Hydrostatics, Hydraulics and Pneumatics: The pressure of fluids is equally diffused and varies as the depth; the surface of a fluid at rest is horizontal; Specific Gravity; a floating body displaces exactly its weight of fluid, and is supported as if by a force equal to its weight pressing upwards at the centre of gravity of the displaced fluid; the Common Pump and the Forcing-Pump; the Barometer; the Air-Pump; the Steam-Engine.
Astronomy: The apparent motion of the heavens round the earth; the apparent motion of the sun through the fixed stars; the phenomena of Eclipses; the Regression of the Planets; proofs of the Copernican system.
Chemistry: Heat: natural and artificial sources, its effects. Expansion; solids, liquids, gases. Thermometer; conduction; radiation; capacity; change of form; liquefaction; steam. Relation of chemical affinity in the voltaic pile; ordinary electricity; its excitement and effects.

8 Extracts from the Regulations and Examination Papers of the London BA Honours Degree in Mathematics and Natural Philosophy, 1839

Candidates will be examined in Algebra, Analytical Geometry; Differential and Integral Calculus; Calculus of Finite Differences; Theory of Probability; Statics and Dynamics; Hydrostatics; Hydraulics; Pneumatics; Heat; Optics; Astronomy.

Science Questions:

Wednesday, June 5th

7. A pendulum oscillates in a medium in which the resistance varies as the velocity; find the time of a small oscillation.
8. A circle of which the plane is vertical is just immersed in a fluid; how may it be divided by a horizontal line into two such parts, that the pressure on them shall be equal?
9. Explain fully the principle of the superposition of small movements in the theory of sound.

Wednesday, June 5th

1. State the ordinary laws of reflection and refraction of light at surfaces of continued curvature. Describe the experiments which establish the truth of such laws. Prove that such laws represent the effects of forces normal to the surfaces on the theory of emitted particles and also the propagation of motion between elastic fluids on the theory of undulations, and in the case of refraction point out the essential differences between both theories relative to the velocity of the refracted rays.
2. Investigate the general formula for the relation of the conjugate foci when reflection takes place from a spherical speculum, and state in what circumstances the image of a body in such a speculum will be inverted.

3. Find the extreme lateral aberration of parallel rays refracted by a double convex lens of equal curvature on both sides.
4. Explain the chromatic dispersion of light on the theory of emission, and investigate a combination of lenses proper to produce achromatism.
5. Find the law for the velocity of the extraordinary refracted beam of light in Iceland spar, according both to the theory of emission and undulations.
6. Give the Newtonian law for the interchange of heat between bodies of different temperatures, and explain the processes by which Dulong and Petit established a more exact law when the difference of temperatures was considerable.
7. A circular ring is subjected at one point to a permanent temperature zero, and at the opposite point, to a permanent temperature 100° (Centigrade); it radiates freely in vacuo and its interior conducting power is to its exterior as a to b; find the permanent temperature of any part of the ring.
8. Give reasons from phenomena for the truth of the Copernican system, more particularly for the diurnal and orbital motions of the earth.
9. Find the latitude of a star of which the right ascension is 301° 20', and the longitude 8° 15'.
10. Explain clearly the causes of precession and nutation.
11. Explain the aberration of light, and show how the velocity of light may be estimated from it; what confirmation may be hence deduced to the undulatory theory of light.
12. How has the origin of aerolites been explained? If supposed to be discharged from lunar volcanoes, what would generally be their apparent direction in our atmosphere.

9 Regulations for the Intermediate BSc Pass Examination 1882

(University of London Calendar 1882)

Candidates will be examined in Pure Mathematics, Mixed Mathematics, Inorganic Chemistry, Experimental Physics, General Biology.

Experimental Physics

Candidates will be expected to show a general acquaintance with the Methods and Apparatus by which the leading principles of Physics, as enumerated below, can be illustrated and applied.

Units of Measurement.

Laws of Motion considered experimentally.

The chief Forces of Nature. The general properties of Solids, Liquids, and Gases. The nature, intensity, and transmission of Fluid pressure in general. The Pressure of Liquids in Equilibrium under the action of Gravity. The Equilibrium of Solids floating or entirely immersed in gravitating Fluids. The Specific Gravities of substances, with the ordinary modes of determining them. Measurement of the Pressure of the Atmosphere and of the Elastic Force of Gases. Diffusion of Liquids and Gases. Definition of Work and Energy; Conservation and Transmission of Energy.

Acoustics

Production and mode of Propagation of Sound. Intensity, Pitch and Quality. Velocity of Sound in Air. Influence of Temperature and Density. Velocity of Sound in other Media. Laws of Reflexion and Refraction. Nature of Musical Sounds. Longitudinal Vibrations of Rods and Columns of Air. Transverse Vibrations of Strings. Variation in their Rate of Vibration by changes in their tension, length, thickness, and substance.

Heat

Definitions of Heat and Temperature. Construction of Instruments for the Measurement of Temperature. Expansion of Solids, Liquids, and Gases under Heat. Change of State; Tension of Vapours; Latent Heat. Radiant Heat, its reflexion, refraction, and absorption. Conduction; definition of Thermal Conductivity. Convection. Specific Heat. Mechanical Equivalent of Heat.

Magnetism

Properties of Magnets. Induction. Magnetic relations of Iron and Steel. Terrestrial Magnetism.

Electricity

Two Electrical States; and their mutual relations. Conduction and Insulation. Induction. Electric Attraction and Repulsion. Distribution and Accumulation of Electricity on Conductors. Electric Discharge. Voltaic Electricity; the various Batteries. Electromotive Force, Strength of Currents, Resistance; Ohm's Law. Heating and Chemical effects of Electric Currents. Action between Currents and Magnets; Electro-Magnetism. Induced Currents; Magneto-Electricity. Thermo-Electricity.

Optics

Laws of Propagation of Light. Measurement of Velocity of Light. Photometry. Laws of Reflexion and Refraction of Light. Reflexion at Plane and at Spherical Surfaces. Refraction at Plane and at Spherical Surfaces. Refraction through Lenses, including the formation of Images. Chromatic Dispersion.

Honours Experimental Physics:

The subjects of the Pass examination treated more fully.

10 Extracts from H.E. Armstrong's 'Notes on the Teaching of General Elementary Science and the Elements of Scientific Method'[3]

The following scheme is drawn up by way of illustration – not so much with the object of laying down a detailed programme but rather in order to show how the subject may be treated.

Object studies

Object Studies – not object lessons, in which the teacher does almost everything – should form the basis of the instruction and every attempt should be made from the beginning to give these a quantitative form.

Measurement of Length and Area – At as early a stage as possible some question should be raised the answer to which may be found by weighing; it is all-important that this should not take the form of a mere exercise but should be regarded by the children as an inquiry. .. As an illustration Drummond's story called The Monkey that would not Kill (Hodder & Stoughton, London, 1898) may be taken. In this a description is given of a troublesome monkey being cast over a cliff into the sea, with a stone tied round his neck, in order to get rid of him; in some way, however, the monkey finds out that under water the stone weighs less than it did on land and that consequently he can lift it, so he picks it up and walks ashore, thus saving his life. Such a story appeals to young children and strikes their imagination; under ordinary circumstances, however, they would probably not question the incident but it is easy to lead them to consider: 'Could such a thing happen?' 'Does a stone weigh less in water than in air?' 'May it not be well to check such a statement, if possible?' The experiment thus suggested being tried and the statement verified, a start is made and it becomes easy to extend the inquiry. 'To what extent does the stone lose in weight?' 'Do all stones lose weight in the same proportion?' 'The loss in weight obviously has something to do with the size of the stone – how is it related to it?' 'Do other things besides stones lose weight in water; if so to what extent?' These and other similar questions come almost as a matter of course and obviously can be answered by experiment. When a variety of objects are weighed in air and water, the children are led to notice that the change in weight has more to do with the size than with the weight of the object – and this renders the determination of bulk necessary. Ultimately it is discovered that the loss in weight is proportional to the volume of the object and equal to the weight of the volume of the water the object displaces. The method of determining volume by ascertaining loss of weight in water is thus discovered. The investigation may then be extended to sea-water and other liquids. The conception of density having been arrived at in the course of such experiments, children may be led to determine the densities of the various materials at their disposal and to express these in grams per cubic decimetre – not as mere relative densities. (The term specific gravity should be altogether avoided.) The opportunities such work affords of making diagrams and models and of setting arithmetical and geometrical exercises should be fully utilised.

Properties of Water ... In order to fix children's attention and to give them a due sense of the importance of the work in which they are engaged, it will probably be found well to centre inquiries around a story such as that of The Three Giants in Stead's penny series of *Books for the Bairns*. The subjects suggested for study by such an allegory are the carrying power of water, as leading up to the understanding of ships; the power of moving water to do work when applied to the water wheel, as leading up to the conception of energy and its measurement; the power of water to do work as steam, as leading up to the conception of heat energy and its measurement and the steam engine. The study of water as ice should be undertaken at the first convenient opportunity ... From data obtained, the expansion of water on becoming ice may be calculated and the consequence of the expansion may be discussed – the bursting of pipes, the breaking up of the soil by frost, .. strong bottles full of water may be placed out of doors during frost. Such experiments live in the memory. Icebergs ... calculations of the extent to which ice is immersed in ordinary and in sea-water when it floats ... The temperatures of melting ice will be another object to study. ...The study of steam ... How hot must water be to make it pass off as steam...? Determination of boiling point... The extent to which water expands on becoming steam...the steam engine... The pressure exerted by steam....Does ordinary gas exert any pressure? The experiment is easily tried and the use of a water gauge made obvious. What pressure can be exerted by a person blowing? A simple gauge is fitted up for the purpose with which the blowing power of each member of a class is tested. Supposing the pressure to be measured is big, what is to be done? Probably some will say, 'Use a bigger' – meaning a wider 'tube': let this be tried. It is then soon discovered that it is entirely a question of height and not of width of column – and a most important hydrostatic principle is discovered.

The use of mercury, as being a very dense liquid, for measuring considerable pressure ... how high a column of liquid can the air pressure support? ... Data are obtained permitting of the determination of the relation between volume and pressure. The results are plotted on squared paper so as to find the law (Boyle's law) ... with the aid of the simplest possible apparatus and without using mercury ... the fluctuations of the barometer. The method of measuring gaseous pressure ... the pressure developed on heating water ... the alterations in the boiling point conditioned by alterations of pressure is then discovered ... the working of the steam engine boiler.

11 'A Nuffield View of Physics', by John Lewis[4]

I first encountered physics as something which appeared in my school timetable when I first entered my secondary school and which most of us spelt with some difficulty. It appeared to be a subject which changed in a remarkable way as the terms went by. One term it was light, later it turned into heat, later still it became magnetism, which seemed to merge in due course into something called electricity. The examination papers included mechanics, but that was left to the mathematics lessons, and also sound, but there never seemed to be time in the overcrowded syllabus to learn much about that. I suppose the predominant feeling left with me was that physics was a conglomerate of topics with very little link between them.

And the teaching? Each topic usually began with a series of definitions, which I rather enjoyed learning by heart as they earned me a lot of marks in every examination I took. In the laboratory there were the pin-sticking processes for finding refractive indices and that immense series of experiments on finding the focal length of lenses; there were all those magnetic fields to be plotted – the A and B position of Gauss – and Hibbert's magnetic balance; and then I first encountered the Wheatstone bridge and the potentiometer, which seemed to dominate most of the rest of my time at school as they were put to more and more ingenious uses. In heat there was Bunsen's ice calorimeter, Hope's apparatus, Joly's differential steam calorimeter, Regnault's hydrometer and then suddenly the name of Searle began to appear – Searle's apparatus for thermal conductivity, Searle's apparatus for measuring Young's modulus and the rigidity modulus of a metal rod.

My laboratory notebook became fuller and fuller as more and more quantities were measured, always written up neatly under the headings Object, Method, Theory, Observations, Calculation, Conclusions. We felt suddenly very modern when we started to work with thermionic valves. We plotted the characteristic curves of diodes and then we progressed to triodes, which gave us a larger series of graphs.

There was of course another aspect of practical work, which was called 'verifying the law'. We happily confirmed that Hooke's law was true because the Laboratory Manual of Instructions told us not to go beyond a certain point for fear of damaging the apparatus. We were good at showing that a plot of current against voltage gave a straight line (with the carefully selected resistance wire provided for us by the apparatus manufacturer) and thereby we 'confirmed Ohm's law'. Straight-line graphs were of course what pleased our teachers and we did our best to provide them.

On the theoretical side there were those great assets for passing examinations, the formulae, all of which we learnt religiously in evenings before the examinations, I never objected at the time because I liked the game, but I wonder now how much physics I learnt. I regret to have to admit that the compartmental view of physics remained throughout my university course and I only have to look again at my thermodynamics notes to see how much I was dominated by formulae rather than physical principles. The practical work still consisted mainly of measuring constants.

It was probably the fact that university physicists were much concerned with measurement in the last half of the 19th century that this should have dominated school physics in the first half of the 20th. Things began to change in the late 1950s when the Association for Science Education (the Science Masters' Association as it then was) advocated the introduction of a little 'modern physics', by which was merely meant a little physics since 1895. It semed only right that the electron should have a place in the syllabus, that X-rays and radio-activity should be mentioned. And shortly after that came the Nuffield programmes.

All would agree that the Nuffield work has had a profound influence on all school physics teaching; nothing will ever be quite the same as the conditions described in the opening of this article.

It is of course the same physics but there have been changes in the teaching which have revealed much more the true nature of physics than was apparent in the old courses.

First and foremost it has brought out the essential unity of the subject. The subject has become a fabric, in which one part of the subject is shown related to another, something studied in one part is shown to have a contribution to the study of another. The old compartmentalization has gone: the concept of energy and the ideas of wave motion, for example, pervade the whole fabric. And that word 'understanding' repeatedly finds expression throughout the teachers' guides. There is a much greater concentration on the understanding of physical concepts and a greater awareness of these unifying concepts. The practical work concentrates likewise on developing an understanding of those concepts rather than on the measurements typical of the past.

Open-ended experiments on the transformer provide an opportunity for imaginative thinking and a search for patterns., which is another characteristic of physics. Even the electric circuit board used in the early years of the O-level course becomes a useful tool by which the pupil is a 'scientist for the day', finding out for himself, within the discipline imposed by the circuit board, the behaviour of elecric circuits and the laws which govern them. Of course it takes longer to let pupils handle the electromagnetic kit themselves than it would if they were shown a demonstration or given precise instructions, but in that way they are learning to think to do real physics. Learning to make sense of phenomena and acquiring a familiarity with them is an important characteristic of many Nuffield experiments. Perhaps the supreme example showing the unity of physics is the development of quantum mechanical ideas at the end of the Nuffield A-level course. In unit 10, the student needs the ideas of crystal structure studied in unit 1, a knowledge of electrons and energy levels from unit 2, the concepts of field and potential from unit 3, the idea of waves and in particular standing waves from unit 4, the Rutherford model of the atom from unit 5, the concept of photons from unit 8 and statistical ideas from unit 9. It is the best teaching I have encountered anywhere for showing the unity of the subject, for showing that physics as a whole makes sense.

Aspects of Nuffield physics

The Nuffield courses also included much to show the role that models play in the development of science. This reveals that the scientist no longer claims to know all the answers: he is a humbler man who realizes that his model is only a model which is useful to help his understanding until his model or his theory is replaced by a better one ....

The courses have tried to show something of how a scientist thinks, how he looks for evidence and how he uses that evidence to modify his model. It was felt we should have done a service to future generations if we could leave with them some flavour of an enquiring mind once they had forgotten all the facts they had learnt at school. There is one part of the Nuffield A-level course which deserves mention as it reflects an important way in which school physics has changed, namely the electronics section. Those characteristic curves have all gone. The electronics is one place where the spirit of engineering is given priority rather than made a subsidiary issue, where the emphasis is on synthesis rather than analysis, where basic units are put together to do a useful job as an amplifier, a multivibrator, a bistable, NOR gate and so on. This is much more in line with the electronics of the world outside the classroom ....

Of course this article is written by an enthusiast who enjoys teaching Nuffield courses. Is all perfect? No. I think Nuffield has failed in one important respect. Much has been done to bring out the essential concepts of the subject and students have been shown something of how scientists think. But the O-level course did little to show the social implications of their physics. Young people are full of idealism, that is one of the reasons why so many good students want to study law or medicine or social sciences because they believe that through them they can do something worth while with their lives by helping others. We have done too little to show the social relevance of our subject, too little to show what a powerful influence our science can have for the future of mankind. And this remains the one big project that needs still to be tackled in the future.

Notes

Introduction

1. T. Martin, *The Royal Institution* (London, The Royal Institution of Great Britain, 1961) p.7.

Chapter 1

1. L. Thorndike, *University Records and Life in the Middle Ages* (New York, Columbia University Press, 1944) p.15.
2. A.C. Crombie, *Robert Grosseteste and the Origins of Experimental Science* (Oxford, Clarendon Press, 1953), pp.61, 84.
3. L. Thorndike, *A History of Magic and Experimental Science*, Vol.II (New York, Macmillan, 1923), p.440ff.
4. *Robert Grosseteste and the Origins of Experimental Science*, p.14.
5. Ibid., p.135.
6. R.T. Gunther, *Early Science in Oxford*, Vol.1. (Oxford, Oxford Historical Society Publications, 1923), p.269.
7. H. Rashdall, *The Universities of Europe in the Middle Ages*, Vol.1. (Oxford, Clarendon Press, 1895) p.437.
8. Foster Watson, *Beginnings of the Teaching of Modern Subjects in England* (London, Sir Isaac Pitman & Sons, 1909), p.356.
9. Venice 1531.
10. *Pantometria* (1521). From this work it would appear that Leonard Digges was conversant with the principles of the telescope.
11. *Dedicatory Epistle to the Stratiocus* (1579).
12. Preface to an English translation of Euclid (1570). Dee also wrote on mechanics, perspective and on 'burning mirrors'.

Chapter 2

1. M. Ornstein, *The Role of the Scientific Societies in the Seventeenth Century* (Chicago, University of Chicago Press, 1928), p.4.
2. W.S. Monroe, *Comenius, the Evangelist of Modern Pedagogy* (London, Heinemann, 1892), p.1.
3. J. Comenius, *The Great Didactic* (London, A. and C. Black, 1896), Chapter XVIII, pp.25, 28.
4. W. Petty, *Advice for the Advancement of Learning* (London, 1648), pp.5–6.
5. John Dury, *Reformed School* (London, 1650) pp.55–9, 66–8.
6. George Snell, *On the Right Teaching of Useful Knowledge* (London, 1649), p.310.
7. John Ward, *The Lives of the Professors of Gresham College* (London, 1749), p.96.

8. Gunther, *Early Science in Oxford*, Vol.1. (Oxford, Oxford Historical Society, Publications, 1923), p.7.
9. John Webster, *Academiarum Examen* (London, 1653), pp.60–78.
10. John Wilkins, Seth Ward, *Vindiciae Academiarum* (Oxford, Leonard Lichfield, 1654), Preface.
11. John Ward, *Vindiciae Academiarum* (Oxford, Leonard Lichfield, 1654), pp.34–41.
12. G. Compayre, *The History of Pedagogy* (Boston, D.C. Heath, 1909), p 191.
13. R.K. Merton, *Science, Technology and Society in Seventeenth Century England.* In *Osiris*, Vol.4. (Bruges, edited George Sarton, 1938), p.450.
14. W.W.R. Ball, *A History of the Teaching of Mathematics at Cambridge* (Cambridge, Cambridge University Press, 1889), p.49.
15. John Wilkins, *The Mathematical and Philosophical Works of John Wilkins* (London, 1708), Preface.
16. Oxford Historical Society, *Collectanea.* First Series, edited C.R.L. Fletcher (Oxford, Oxford Historical Society Publications, 1885), p.317.
17. P.H. Osmond, *Isaac Barrow* (London, Society for the Promotion of Christian Knowledge, 1844), p.119.
18. Oxford Historical Society, *Collectanea*, p.32.
19. Gunther, *Early Science in Oxford,* Vol.1. p.199.
20. W. Stukeley, *Memoirs of Sir Isaac Newton's Life* (London, Taylor and Francis, 1752), p.10.
21. J.E.B. Mayor, *Cambridge under Queen Anne* (Cambridge, Cambridge Antiquarian Society, 1911), p.457.
22. Ball, *A History of the Teaching of Mathematics at Cambridge*, p.86.
23. John Green, *The Academic: or, a Disputation on the University of Cambridge* (London, C. Say, 1750), pp.23–5.
24. D.A. Winstanley, *Unreformed Cambridge* (Cambridge, Cambridge University Press, 1935), p.151.
25. Mary Milner, *The Life of Isaac Milner* (London, 1842), p.32.
26. Ball, *A History of the Teaching of Mathematics at Cambridge,* p.104.
27. J. Jebb, *Remarks upon the Present Mode of Education in the University of Cambridge* (Cambridge, 1773), pp.21–3.
28. N. Hans, *New Trends in Education in the Eighteenth Century* (London, Routledge & Kegan Paul, 1951) p.54.
29. *Observations on the Present State of the English Universities* (1759), pp.17–21.
30. Vicesimus Knox, *Liberal Education* (London, Charles Dilly, 1781), p.162.
31. Christopher Wordsworth, *Scholae Academicae* (Cambridge, 1877), pp.10–11.
32. *The Scots Magazine* (Edinburgh, August 1771), pp.371–2.
33. D. Murray, *Memories of the Old College of Glasgow* (Glasgow, Maclehose, Jackson & Co., 1927), p.111.
34. James Muir, *John Anderson* (Glasgow, John Smith & Son, 1950), pp.98–9.
35. Gunther, *Early Science in Oxford,* Vol.1. p.9.
36. E. Halley, *Miscellania Curiosa* (London, 1705), Advertisement at end of Vol.1.
37. Mrs Bathsua Makin, *Essay to Revive the Ancient Education of Gentlewomen* (London, 1673).

Chapter 3

1. G. Turnbull, *Observations upon Liberal Education* (London, A. Millar, 1742), p.195.
2. James Burgh, *Thoughts on Education* (London, 1747), p.11.
3. Anonymous, *Education of Young Students in all its Branches* (London, 1752), p.18.
4. David Williams, *Treatise on Education* (London, 1774), p.251.
5. Dr Joseph Priestley, *Memoirs of Dr Joseph Priestley* (London, J Johnson, 1806), p.9.
6. Irene Parker, *Dissenting Academies in England* (Cambridge, Cambridge University Press, 1914).
7. J. Toulmin, *An Historical View of the State of the Protestant Dissenters in England* (Bath, 1814), pp.227–9.
8. *Dictionary of National Biography.*
9. Samuel Wesley, *A Letter from a Country Divine.. on the Education of Dissenters* (London, R. Clavel, 1703).
10. M. Henry, *Life of Dr Samuel Benion* (London, 1709), pp.67–78.
11. W. Wilson, *Dissenting Churches*, Vol.II (London, 1808), p.75.
12. J. Orton, *Memoirs of Philip Doddridge* (Salop, J. Cotton and J. Eddowes, 1766), p.91.
13. John Williams, *Memoirs of T. Belsham* (London, 1833), p.226.
14. Vicesimus Knox, *Personal Nobility* (London, Charles Dilly, 1793), pp.128–31.
15. Priestley, *Memoirs*, p.41.
16. *Monthly Repository*, Vol. VIII (London, 1814), p.4.
17. G. Howson, *A History of Mathematical Education in England* (Cambridge, Cambridge University Press, 1982).
18. Joseph Priestley, *Experiments and Observations relating to Various Branches of Natural Philosophy* (London, 1779), pp.ix–x.
19. *Monthly Repository*, Vol.V, (London, 1810), p.218.
20. John Horsley, *A Brief and General Account of the Principles of Statics, Mechanics, Hydrostatics and Pneumatics* (Glasgow, Andrew Stalker, 1731).
21. John Colet Ryland, *Address to the Ingenious Youth of Great Britain* (London, 1792), pp.121–36.
22. H. McLachlan, *English Education under the Test Acts* (Manchester, University of Manchester, 1931), p.262.
23. Hans, *New Trends in Education in the Eighteenth Century*, p.68.
24. B. Donne, *Mathematical Essays* (London, W. Johnson & B. Law, 1769).
25. Henry Clarke, *The School Candidates*, edited J.E. Bailey (Manchester, T.J. Day, 1877), p.xxxiii.
26. Henry Clarke, *Seaman's Desiderata* (Bristol, Richard Edwards, 1800).
27. J.B. Florian, *Elementary Course of the Sciences and Philosophy* (London, 1806).
28. J. Grant, *History of the Burgh and Parish Schools of Scotland*, Vol.1 (Edinburgh, 1876), p.402.
29. Ibid., p.118.
30. Sir John Sinclair, *Statistical Account of Scotland* (Edinburgh, 1791–99), Vol.X, p.256; Vol.XII, p.38.
31. Her school was at one time Bryan House, Blackheath, at another 27 Lower

Cadogan Place, London, and lastly at Margate.
32. P. Gosden, *How They Were Taught*, (Oxford, Blackwell, 1969), p.149.

Chapter 4

1. James Muir, *John Anderson* (Glasgow, John Smith & Son, 1950), p.25.
2. T. Webster, *A Manual for Mechanics' Institutions* (London, Society for the Diffusion of Useful Knowledge, 1839), p.26.
3. R. Detrosier, *An Address Delivered to the New Mechanics' Institution, Pool Street, Manchester, December 30, 1829* (Manchester, T. Forrest, 1829), p.4.
4. *Leeds Mechanics' Institution, Annual Report* (Leeds, 1856), p.12.
5. Thomas Coates, *Report of the State of Literary, Scientific and Mechanics' Institutions in England* (London, 1841), p.56.
6. *Liverpool Mechanics' Institution, Report* (Liverpool 1832).
7. J.W. Hudson, *A History of Adult Education* (London, 1851), p.52.
8. *Select Committee on Scientific Instruction, Minutes of Evidence* (Professor Tyndall – London, Parliamentary Papers, 1868), p.20.
9. *Coates Report of the State of Literary, Scientific and Mechanics' Institutions in England* (London, 1841), Appendix IV.

Chapter 5

1. H. Bence Jones, *The Life and Letters of Faraday*, Vol.I (London, Longmans, Green & Co., 1870), pp.11–14.
2. T. Martin, *The Royal Institution* (London, The Royal Institution of Great Britain, 1961), p.22.
3. Bence Jones, *The Life and Letters of Faraday*, Vol.II, p.228.
4. Ibid., p.451.
5. John Tyndall, *Lessons in Electricity* (London, Longmans, Green & Co., 1876), Preface.
6. Martin, *The Royal Institution*, p.56.

Chapter 6

1. Thomas Wyse, *Education Reform* (London, 1836), p.162.
2. Alfred Bourne, *Royal Commission on Scientific Instruction and the Advancement of Science, Minutes of Evidence, Second Report* (London, Parliamentary Papers, 1872), p.562.
3. William Lovett, *Address to the Working Classes* (London, London Trades Combination Committee, 1837).
4. W.Jolly, *Education: Its Principles and Practice as developed by George Combe* (London, Macmillan & Co. 1879), p.233.
5. *Williams Secular School, First Annual Report* (Edinburgh, 1850), p.16.
6. Ibid., p.14.

7. *Lectures in Connection with the Educational Exhibition of the Society of Arts, Manufactures and Commerce* (London, Royal Society of Arts, Manufacturers and Commerce, 1854), p.272.
8. Richard Dawes, *Hints on an Improved and Self-Paying System of National Education*, Fourth Edition (London, Groombridge & Sons, 1850), p.17.
9. Ibid., pp.17–18.
10. D. Layton, *Science for the People* (London, Allen & Unwin, 1973), pp.42–3.
11. A. Sutcliffe, *Students' Laboratories in England, School Science Review,* Vol. II. (London, Science Masters' Association, 1929), p.87.
12. Bence Jones, *The Life and Letters of Faraday*, Vol.II, p.393.
13. Ibid., p.209.
14. R. Hunt, 'On Familiar Methods of Instruction in Science' (Lecture given at Educational Exhibition of the Society of Arts, 1854), p.178.
15. J.P. Kay-Shuttleworth, *Four Periods of Public Education* (London, 1862), p.343.
16. *Her Majesty's Inspectors, Reports on Normal Schools* (London, Parliamentary Papers, XXXIX, 1851), p.271.
17. R.L. Archer, *Secondary Education in the Nineteenth Century* (Cambridge, Cambridge University Press, 1937), p.344.

Chapter 7

1. Augustus de Morgan, 'On the Study of Natural Philosphy' in *The Schoolmaster: Essays on Practical Education*,Vol.II, p.233 (London, 1836).
2. *Quarterly Journal of Education*, Vol.III (London, Society for the Diffusion of Useful Knowledge, 1832), p.45 footnote.
3. Thomas Wyse, *Education Reform* (London, 1836), pp.160–2.
4. J. Booth, Education and Educational Institutions (London, J.W. Parker, 1846), p.103.
5. William Robinson, *History and Antiquities of Hackney*, Vol.II (London, J.B. Nichols & Son, 1842), p.229.
6. Council of the Society of Arts, *Report on Industrial Education* (London, 1853), p.183.
7. A.A. Mumford, *The Manchester Grammar School* (London, Longmans & Co., 1919), p.272.
8. Henry Malden, *On the Introduction of the Natural Sciences into General Education* (London, October 1838), p.8.
9. Ibid., pp.16–17.
10. See N. Carlisle's *Concise Description of the Grammar Schools in England and Wales* (London, Baldwin, Cradock, and Joy, 1818), for a list of schools where this more liberal curriculum was permitted.
11. J.W. Adamson, *History of Education 1792–1902* (Cambridge, University Press, 1919), p.49.
12. E.L. Greenberg, *The Contribution made by the Private Academies in the First Half of the Nineteenth Century to Modern Curriculum and Methods* (London University, MA Thesis, 1952), p.120.

13. William Johnstone, *Results of Experience in the practice of Instruction* (London, 1818), pp.5, 62.
14. Matthew Davenport Hill, *Sketch of the System of Education at Bruce Castle, Tottenham* (London, 1833), p.37.
15. *Royal Commission on Scientific Instruction and the Advancement of Science* (London, Parliamentary Papers, 1876), Appendix, University College School p.116.
16. *Brief Account of the City of London School* (London, Taylor, 1841). [British Library 1034m26.]
17. John Collis Nesbitt, *Prospectus of the Classical, Commercial, Mathematical and Scientific Academy at 38, Kensington Lane, Lambeth* (London, 1846), p.7.
18. Ibid., p.14.
19. Mumford, *The Manchester Grammar School*, p.270.
20. Bristol College, *Outlines of the Plan of Education to be pursued in the Bristol College* (Bristol, 1830), p.6.
21. Greenberg, *The Contribution made by the Private Academies*, p.91.
22. Ibid., p.143.
23. John Shoveller, *Essay on the most Effectual Methods of Advancing Youth in Scholastic Education* (Portsea, 1815), p.64.
24. *Lectures in Connection with the Educational Exhibition*, p.44.
25. Ibid.
26. Ibid.
27. Henry Cockburn, *Journal* (Edinburgh, Edmondston and Douglas, 1874), p.69.
28. Mr M'Laren, *Address to the Inhabitants of Dundee* (Dundee, 1834), p.36.

Chapter 8

1. T.H. Huxley, *Science and Education, Collected Essays*, Vol.III. (London, Macmillan & Co., 1893), pp.121–31.
2. *Papers relating to Proposals for Establishing Colleges of Arts and Manufactures for the Better Instruction of the Industrial Classes* (London, 1851), p.14.
3. Lyon Playfair, *Lectures on the Results of the Great Exhibition of 1851* (London, 1851), p.197.
4. John Tyndall, *Correspondence*, Vol.3., Letter 982.
5. Any School, or part of a school, that taught science and so earned grants from the Science and Art Department was called a Science School. A 'School' of Science was not really a school at all, but usually a portion of a school, secondary or elementary, organized to receive South Kensington grants.
6. Huxley, *Science and Education*, p.132.
7. *Department of Science and Art, Syllabus 1861* (London, 1861), p.34.
8. *Royal Commission on Scientific Instruction and the Advancement of Science, Second Report* (London, Parliamentary Papers, 1872), p.xii.
9. Ibid., p.xxvi.
10. *Select Committee on Scientific Instruction* (London, Parliamentary Papers, 1872), p.415.
11. *Schools Inquiry Commission, Minutes of Evidence* xxvii, p.3. (London, Parliamentary Papers, 1865), p.643.

12. *Royal Commission on Scientific Instruction ... Second Report, Minutes of Evidence* (London, Parliamentary Papers, 1872), pp.401–2.
13. T.H. Huxley, *'A Liberal Education and Where to Find It', Collected Essays*, Vol.III (London, Macmillan & Co., 1868), p.151–2.
14. *Royal Commisson on Scientific Instruction ...* , Appendix XVII 1872, p.53.
15. *Royal Commission on Scientific Instruction ... Second Report, Minutes of Evidence*. 1872. p.25.
16. Ibid., *Minutes of Evidence* (3 July 1868), p.401.
17. Ibid., Appendix XVII, p.53.
18. Ibid., *Second Report*, Vol.1. 1872.
19. Archer, *Secondary Education in the Nineteenth Century* (Cambridge, Cambridge University Press, 1937), p.420.
20. T.H. Huxley, *Address on Behalf of the National Association for the Promotion of Technical Education* (London, 1887).
21. Huxley, *'A Liberal Education'*, pp.153, 163.
22. E.W. Jenkins, *From Armstrong to Nuffield* (London, John Murray, 1979), p.20.

Chapter 9

1. Huxley, *'A Liberal Education'*.
2. H.E. Armstrong, *The Teaching of Scientific Method and Other Papers on Education*, Second Edition (London, Macmillan & Co., 1903), pp.xiv–xxi.
3. A.F. Leach, *History of Winchester College* (London, Duckworth & Co., 1899), p.465.
4. Gunther, *Early Science in Oxford*, Vol.III, p.81.
5. *Report of the Public Schools Commission,* Vol.IV (London, Parliamentary Papers, 1864), p.337.
6. *Royal Commission on Scientific Instruction and the Advancement of Science*, Appendix V (London, Parliamentary Papers, 1872).
7. *Report of the Public Schools Commission* (London, 1864), p.278.
8. *Royal Commission on Scientific Instruction ... Sixth Report*, Appendix (London, Parliamentary Papers, January 1876), p.107.
9. A. Sutcliffe, *'Student Laboratories in England', School Science Review*, Vol.II. 1929. p.80.
10. *Report of the Public Schools Commission, Evidence of Mr Wilson* (London, Parliamentary Papers, 1864), p.252.
11. *Reports from the Commissioners, Public Schools and Colleges*, Vol.XX. (London, Parliamentary Papers, 1864), p.86. The subjects taught included astronomy, heat, light, electricity, chemistry, coal-mining, telegraphy, photography and even spectrum analysis.
12. Leach, *History of Winchester College*, p.465.
13. *Report of the Public Schools Commission*, Vol.1. 1864, pp.145–6.
14. Ibid.
15. Leach, *History of Winchester College*, pp.468–70.
16. *Report of the Public Schools Commission*, Vol.1. p.311.
17. Ibid., p.55.
18. Ibid., p.98.

19. Ibid., *General Report, First Part*, 1864, pp.32–3.
20. Ibid., *Report*, Vol.1. 1864, p.115.
21. Ibid., *Minutes of Evidence* 1864, p.278.
22. Ibid., *Report* 1864, p.101.
23. *Royal Commission on Scientific Instruction ... Sixth Report*, Appendix, 1876, pp.106–9.
24. Ibid, p.108.
25. J.M. Wilson, *Natural Science in Public Schools*, Paper read to the British Association for the Advancement of Science, September 1866.
26. *Royal Commission on Scientific Instruction ... Sixth Report*, Appendix, 1876, p.159.
27. *Report from a Committee appointed by the Council of the British Association for the Advancement of Science to consider the Best Means for Promoting Scientific Education in Schools*, Appendix A (Dundee, llth March, 1868), pp.12–13.
28. *Royal Commission on Scientific Instruction ... Sixth Report*, Appendix, 1876, p.109.
29. *Royal Commission on Scientific Instruction ... Sixth Report*, Appendix, 1876, pp.188–9.
30. *Schools Inquiry Commission, Minutes of Evidence*, xxvii, 3 (London, Parliamentary Papers, 1865), p.174.
31. Leach, *History of Winchester College*, p.487.
32. Ibid., p.533.
33. *Royal Commission on Scientific Instruction ... Sixth Report*, Appendix 1876, p.141.
34. Ibid., p.131.

Chapter 10

1. *Stonyhurst College Archives*, MS B/ii/15.
2. *Stonyhurst College Magazine*, ix (Stonyhurst, 1903), p.23.
3. *Royal Commission on Scientific Instruction ... Sixth Report*, Appendix, 1876, pp.121–2.
4. Ibid., p.161.
5. *Schools Inquiry Commission. Evidence of Cheltenham College* (London, Parliamentary Papers, 1865).
6. *Nature*, Vol.XXIV (London, 1881), p.543.
7. *Nature* (London, 13 October 1881), p.557.
8. *Royal Commission on Scientific Instruction ... Sixth Report*, Appendix, 1876. pp.182–3.

Chapter 11

1. *Schools Inquiry Commission, Report*, Vol.I (London, Parliamentary Papers, 1868), p.502.
2. Ibid., Vol.XXVIII (1875) p.103.

3. Ibid., Vol.IX, p.237.
4. *Westminster Budget* (London, 15 December 1893).
5. *Schools Inquiry Commission*, Vol.1. 1864, p.148.
6. Ibid., *Analysis of Evidence*, Vol.II 1867/8, p.309.
7. W. Hall, *The Teaching of Science in England during the Latter Half of the Nineteenth Century and Huxley's Influence on it* (Sheffield University, MA Thesis, 1931).
8. *Royal Commission on Scientific Instruction ... Sixth Report*, Vol.XXVIII, folio, p.2.
9. Ibid., Appendix, 1875, Section 20.
10. Ibid., Vol. XXVIII, folio.
11. D.M. Turner, *The History of Science Teaching in England* (University of London, PhD Thesis, 1928), p.94.
12. Archer, *Secondary Education*, p.141.
13. *Royal Commission on Scientific Instruction ... Sixth Report*, p.60.
14. *Schools Inquiry Commission*, Vol.I, 1868, p.224.
15. *Royal Commission on Scientific Instruction ... Sixth Report*, p.8.
16. Ibid., *General Remarks on Laboratories for Science Schools.*
17. Ibid., Appendices 4, 5.

Chapter 12

1. *Public Schools Commission, Minutes of Evidence*, p.363.

Chapter 13

1. H.E. Armstrong, *The Teaching of Scientific Method and Other Papers on Education*, second edition (London, Macmillan, 1925), pp.204.
2. Ibid., pp.168, 198.
3. Ibid., pp.57–9.
4. Ibid., p.240.
5. Ibid., p.64.
6. Ibid., p.70.
7. *Nature*, 28 January 1904, p.289.
8. H.E. Armstrong, *Address to the Educational Sector of the British Association for the Advancement of Science* (Belfast, 1902).
9. H.E. Armstrong, 'The Teaching of Scientific Method', in the *Educational Times* (London, May 1891).
10. Armstrong, *The Teaching of Scientific Method and other Papers in Education*, p.229.
11. H.E. Armstrong, 'An Appeal to Headmasters', in *Journal of Education* (London, January 1901), p.20.
12. Ibid.
13. Quoted in Armstrong, 'An Appeal to Headmasters', p.20.
14. Armstrong, 'An Appeal to Headmasters', p.20.
15. H. Bernard, *English Pedagogy*, Vol.II (Philadelphia, J.P. Lippincott & Co., 1876), p.478.

16. Ibid.
17. J.M. Wilson, 'On the Teaching of Natural Science in Schools', in F.W. Farrar (ed.), *Essays on a Liberal Education* (London, Macmillan, 1867), p.272.
18. H. Bernard, *English Pedagogy*, Vol II, p.478.
19. H.E. Armstrong, Huxley Memorial Lecture, 4 May 1933, in Armstrong, *Papers on Education*.
20. Ibid., pp.72–3.
21. *British Association for the Advancement of Science. Report* (London 1889), pp.228–9.
22. *The Educational Times*, 44 (London, 1891), p.228.
23. In F. Spencer, *Chapters on the Aims and Practice of Teaching* (Cambridge, Cambridge University Press, 1897), pp.214–15.
24. Ibid.,pp 211–12.
25. Armstrong, *Papers on Education* pp.193, 195.
26. *Nature* (London, 10 October 1901), p.591.

Chapter 14

1. *Schools Inquiry Commission, Minutes of Evidence*, p.402.
2. W. Tuckwell, *Reminiscences of Oxford* (London, Cassell & Co., 1900), p.41.
3. Gunther, *Early Science in Oxford*, Vol.1., pp.75–6.
4. Baden Powell, *The Present State and Future Prospects of Mathematical and Physical Studies in the University of Oxford* (Oxford, 1832), pp.8–10, 21.
5. C. Daubeny, *Supplement to the Introduction to the Atomic Theory* (Oxford, J.H. Parker, 1840), Preface.
6. Sir William Hamilton, *Discussions* (Edinburgh, 1852), p.821.
7. W.D. Macray, *Annals of the Bodleian Library* (London, Oxford, 1868).
8. *Nature*, Vol.1, 1869, p.26.
9. *Nature*, Vol.50, 1894, p.344.
10. *Royal Commission on Scientific Instruction ... Third Report, Minutes of Evidence* (London, Parliamentary Papers, 1873), p.187.
11. Ibid.
12. Ibid., Vol.XXXI. p.xvii.
13. Ibid.
14. *Quarterly Journal of Education*, Vol.X, p.149.
15. W. Whewell, *On the Principles of English University Education* (London, 1837), p.72.
16. *Report of Her Majesty's Commissioners appointed to inquire into the University of Cambridge and Colleges of Cambridge* (London, Parliamentary Papers, 1852), p.65.
17. *Report of Her Majesty's Commissioners ...*, p.94.
18. L. Campbell and W. Garnett, *The Life of James Clerk Maxwell* (London, Macmillan & Co., 1882), p.86.
19. Ibid., p.115.
20. Ibid., pp.117–20.
21. Ibid., p.106.

22. R.T. Glazebrook, *James Clerk Maxwell and Modern Physics* (London, Cassell & Co. 1896), p.25.
23. Ibid., p.32.
24. Campbell and Garnett, *The Life of James Clerk Maxwell*, p.216.
25. Ibid., pp.164–5.
26. *Royal Commission on Scientific Instruction ... Third Report. Minutes of Evidence*, p.275.
27. University of Cambridge, *A History of the Cavendish Laboratory* (London, Longman Green & Co., 1910), pp.79–81.
28. *Monthly Review* (London, January 1829), p.5.
29. *Royal Commission on Scientific Instruction ... Fifth Report*, Vol.XXII (1874), p.11.
30. Ibid., p.5.
31. Ibid., p.6.
32. *A History of the Cavendish Laboratory*, p.2.
33. Campbell and Garnett, *The Life of James Clerk Maxwell*, 1882, p.318.
34. F. Cajori, *A History of Physics* (New York, Dover Publications, 1922).
35. *Royal Commission on Scientific Instruction ... Minutes of Evidence* (Professor W.G. Adams), p.448ff.
36. *A History of the Cavendish Laboratory*, p.77.
37. A. Grant, The *Story of the University of Edinburgh* (London 1884), Vol.2, p.115.
38. *Royal Commission on Scientific Instruction ... Vol.XXII Fourth/Fifth Report*, 1874, pp.12–15.
39. A. Grant, *The Story of the University of Edinburgh*, p.356.
40. D. Murray, *Memories of the Old College of Glasgow* (Glasgow, Maclehose, Jackson & Co., 1927), pp.119–24.
41. D. Murray, *Lord Kelvin as Professor in the Old College of Glasgow* (Glasgow, Maclehose, Jackson & Co., 1924), pp.1–11.
42. D.J.S. Sutherland, *The History of the Teaching of Science in Scottish Schools* (St Andrews University, PhD Thesis, 1939), p.266.
43. James Coutts, *History of the University of Glasgow* (Glasgow, James Maclehose & Sons, 1909), p.388.
44. C.W. Everitt, *James Clerk Maxwell* (New York, Charles Scribner's Sons, 1975), pp.52–3.

Chapter 15

1. Glazebrook, *James Clerk Maxwell and Modern Physics*, p.67.
2. *A History of the Cavendish Laboratory*, p.34.
3. Ibid., p.19.
4. Everitt, *James Clerk Maxwell*, p.396.
5. *A History of the Cavendish Laboratory*, p.37.
6. Ibid., pp.31–9.
7. Ibid., p.34.
8. Everitt, *James Clerk Maxwell*, p.370.
9. A.E. Munby, *Laboratories – Their Planning and Fittings*, Introduction by Sir A.E. Shipley, p.xiii (London, G Bell & Sons, 1921).
10. Everitt, *James Clerk Maxwell*, p.357.

11. *A History of the Cavendish Laboratory*, p.43.
12. Ibid., pp.47–9.
13. Ibid., p.81.
14. J.G. Crowther, *The Cavendish Laboratory 1874–1974* (London, Macmillan, 1974) p.120.
15. Ibid., p.156.
16. Ibid., p.167.
17. Ibid., p.168.
18. Ibid., p.241.

Chapter 16

1. Jenkins, *From Armstrong to Nuffield* (London, Murray, 1979), p.51.
2. J.J. Thomson (Chairman), *Report: Natural Science in Education* (London, Parliamentary Papers, 1918), p.279.
3. *Consultative Committee on Examinations in Secondary Schools* (London, Parliamentary Papers, 1911), p.67.
4. Mr Frewin was author of *Experimental Science*, a textbook that incorporated a more modern approach to teaching science.
5. *Hadow Report, The Education of the Adolescent* (London, Report of Consultative Committee, Board of Education, 1926), p.223.
6. Jenkins, *From Armstrong to Nuffield*, pp.92–8.
7. John Lewis, 'A Nuffield View of Physics', in *Physics Education*, Vol.12, No.2 (March 1977), p.72.

Appendices

1. From the Minutes of Committee of Council on Education 1856–7, p.1131.
2. Royal Commission on Scientific Instruction and the Advancement of Science, Appendix to Sixth Report, 1875, XXVIII folio, p.68.
3. Being extracts from the Memorandum of the Departmental Committee on the Training College Course of Instruction, 1901, H.E. Armstrong, *Papers on Education*, pp.428–44.
4. In *Physics Education*, Vol.12, No.2 (March 1977), pp.70–2.

Select Bibliography

Adamson, J.W. *History of Education 1792–1902*, Cambridge, Cambridge University Press, 1919.

Archer, R.L. *Secondary Education in the Nineteenth Century*, Cambridge, Cambridge University Press, 1937.

Crowther, J.G. *The Cavendish Laboratory 1874–1974*, London, Macmillan, 1974.

Everitt, C.W. *James Clerk Maxwell*, New York, Charles Scribners' Sons, 1975.

Gosden, P. *How They Were Taught*, Oxford, Blackwell, 1969.

Hans, N. *New Trends in Education in the Eighteenth Century*, London, Routledge & Kegan Paul, 1951.

Howson, G. *A History of Mathematical Education in England*, Cambridge, Cambridge University Press, 1982.

Jenkins, E.W. *From Armstrong to Nuffield*, London, John Murray, 1979.

Jones, H. Bence *The Life and Letters of Faraday*, London, Longmans, Green & Co., 1870.

Layton, D. *Science for the People*, London, Allen & Unwin, 1973.

Martin, T. *The Royal Institution*, London, the Royal Institution of Great Britain, 1961.

Ornstein, M. *The Role of the Scientific Societies in the Seventeenth Century*, Chicago, University of Chicago Press, 1928.

Parker, Irene *Dissenting Academies in England*, Cambridge, Cambridge University Press, 1914.

Watson, Foster *Beginnings of the Teaching of Modern Subjects in England*, London, Sir Isaac Pitman & Sons, 1909.

Index